Modern
Potting Composts

Modern Potting Composts

A Manual on the Preparation and Use of Growing Media for Pot Plants

A. C. BUNT

N.D.H.(Hons), M.I. Biol.

Glasshouse Crops Research Institute,
Littlehampton, Sussex

THE PENNSYLVANIA STATE UNIVERSITY PRESS
University Park and London

First published in 1976

This book is copyright under the Berne Convention. All rights are reserved. Apart from any fair dealing for the purpose of private study, research, criticism or review, as permitted under the Copyright Act, 1956, no part of this publication may be reproduced, stored in a retrieval system, or transmitted, in any form or by any means, electronic, electrical, chemical, mechanical, optical, photocopying, recording or otherwise, without the prior permission of the copyright owner. Enquiries should be addressed to the publishers.

 © George Allen & Unwin Ltd. 1976

Published in the United States by
The Pennsylvania State University Press

ISBN 0-271-01221-8

LC 75-42969

Printed in Great Britain
in 10 on 12pt 'Monophoto' Times
by Page Bros (Norwich) Ltd, Norwich

NOTE TO READERS

To assist the reader in the presentation of information, trade names of products have sometimes been used. This constitutes neither endorsement of named products nor criticism of those unnamed.

Whilst every attempt has been made to check the accuracy of the various compost formulae, no liability can be assumed following their use.

Preface

The last two decades have seen rapid advances in the technology used to produce pot plants. Glasshouses designed and orientated to give maximum light transmission, fully automatic heating and ventilating systems, carbon dioxide enrichment of the atmosphere, controlled photoperiods using automatic blackouts and incandescent lamps which enable plants such as chrysanthemum to be flowered at any time of the year, mist propagation techniques, chemical growth regulators which control the height of plants, automatic watering and feeding systems, etc.: these are only some of the developments which have transformed pot plant culture.

There have also been many changes in the composts and systems used to grow the plants. Mineral soils, which formed the basis of the John Innes Composts, are now either too expensive or too difficult to obtain in suitable quality and sufficient quantity. Consequently the grower has been forced to seek other materials such as peat, perlite, vermiculite, plastic foam, shredded bark, etc. New types of fertilisers, new methods of heat sterilisation and new chemical sterilising agents are also being used.

As with many industrial processes, an alteration to improve one part of a process often means that alterations to other parts of the process are required in order to make the whole operation successful; so too with the new composts. By changing the bulky materials from which the compost is made, a different emphasis must be given to the type and quantity of base fertilisers used, and also to the watering and liquid feeding. Composts made with these new materials give results that are equal or even superior to those obtained from the traditional composts, *providing that their individual characteristics and requirements are understood.* The use of these new composts should be regarded as *a new system of growing,* rather than a simple change of compost.

The purpose of this book is to provide horticulturists, including students, growers, advisory officers and those who simply grow plants for pleasure, with information on the characteristics of the newer materials, how they can be used to make composts and for the subsequent nutrition of the plants. Much of the information given is based on previously unpublished studies made by the author at the Glasshouse Crops Research Institute over the last fifteen years.

ACKNOWLEDGEMENTS

It is a pleasure to acknowledge the helpful comment and advice received from the Director, Dr D. Rudd-Jones, and colleagues at the GCRI, especially to Dr B. Acock, Mr P. Adams, Dr M. C. Powell, Mr G. F. Sheard and Dr G. W. Winsor for reading parts of the manuscript; also to Mr G. A. Wadsworth, ADAS. I am also grateful to Professors J. W. Boodley and J. G. Seeley of Cornell University, and to Professors J. K. Rathmell Jr and J. W. White of Pennsylvania State University, for advice on points regarding certain USA practices.

Dr R. Arnold Bik (Netherlands), Dr R. Gabriels (Belgium), Professor A. Klougart (Denmark), Professor F. Penningsfeld (West Germany), Professor V. Puustjärvi (Finland) and Mr J. C. R. Seager (Ireland) kindly provided information on the formulation of composts in their respective countries.

Grateful acknowledgements are also given to the publishers and the authors for permission to reproduce Figures 2.2, 4.6 and 4.9.

Contents

	page
PREFACE	9
1 WHY CHANGE?	15
1.1 *Loam composts*	18
1.2 *Loamless composts*	18
2 ALTERNATIVE MATERIALS	19
2.1 *Peat*	19
2.2 *Other organic materials*	29
2.3 *Mineral materials*	36
2.4 *Plastics*	40
3 PHYSICAL ASPECTS	43
3.1 *Physical terminology*	43
3.2 *Physical requirements of composts*	45
3.3 *Energy concept of water in composts*	46
3.4 *Water absorption and release by composts*	50
3.5 *Formulation of composts: physical principles*	53
4 PRINCIPLES OF NUTRITION	60
4.1 *Cation exchange capacity*	60
4.2 *Anion exchange capacity*	63
4.3 *Availability of nutrients: loam v. loamless composts*	63
4.4 *Nutrient uptake by the plant*	67
4.5 *Acidity (pH)*	70
4.6 *Lime requirement*	73
4.7 *Soluble salts*	74
5 NITROGEN	82
5.1 *Nitrogen and pot plants*	83
5.2 *Forms of mineral nitrogen*	86
5.3 *Slow release fertilisers*	91
5.4 *Choice of fertiliser type*	98
5.5 *Nitrogen and peat*	103
6 OTHER MACRO-ELEMENTS	107
6.1 *Phosphorus*	108
6.2 *Potassium*	114
6.3 *Calcium*	116

12 Contents

		page
6.4	Magnesium	119
6.5	Sulphur	122
6.6	Mineral soil and peat comparison	122
6.7	Nutrient and environment interactions	123
6.8	Fertiliser analysis and salt index	126
6.9	Plant mineral levels	128

7 MICRO-ELEMENTS — 130

7.1	Boron	131
7.2	Copper	135
7.3	Manganese	136
7.4	Molybdenum	137
7.5	Iron	138
7.6	Zinc	141
7.7	Chloride	141
7.8	Aluminium	142
7.9	Fritted micro-elements	143
7.10	Chelated micro-elements	145
7.11	Other sources	146
7.12	Micro-element availability	148

8 COMPOST FORMULATION AND PREPARATION — 150

8.1	Historical	150
8.2	Denmark	151
8.3	Finland	152
8.4	Germany	153
8.5	Ireland	154
8.6	Netherlands	155
8.7	United Kingdom	155
8.8	United States of America	158
8.9	Sawdust and bark composts	162
8.10	Azalea composts	166
8.11	Proprietary formulations	169
8.12	Compost preparation	169

9 LIQUID FEEDING — 174

9.1	Importance of liquid feeding	174
9.2	Formulating liquid feeds	180
9.3	Practical aspects of feeding	192
9.4	Diluting equipment	197
9.5	Quality of irrigation water	203

10 IRRIGATION SYSTEMS — 205

10.1	Drip systems	205
10.2	Capillary watering	208
10.3	Flooded benches	213

11	**JOHN INNES COMPOSTS**	page	215
	11.1 *Formulation*		215
	11.2 *Compost ingredients: loam*		216
	11.3 *Peat*		219
	11.4 *Sand*		219
	11.5 *Sterilisation*		220
	11.6 *Characteristics and use*		221
	11.7 *Composts for calcifuge plants (JIS(A))*		227
12	**HEAT STERILISATION**		229
	12.1 *Thermal deathpoints*		229
	12.2 *Methods of heat sterilisation*		231
	12.3 *Steam*		231
	12.4 *Steam–air mixtures*		235
	12.5 *Flame pasteuriser*		238
	12.6 *Electrical sterilisers*		239
	12.7 *Other methods*		241
	12.8 *Chemistry of heat sterilisation*		241
	12.9 *Rules for heat sterilisation*		245
13	**CHEMICAL STERILISATION**		247
	13.1 *Soil fumigants*		247
	13.2 *Fungicides*		249
	13.3 *Insecticides*		251
14	**PLANT CONTAINERS**		252
	14.1 *Clay v. plastic pots*		252
	14.2 *Paper and peat pots*		259

APPENDICES		260
1	*Metric conversions*	260
2	*Imperial and US capacity measures*	261
3	*Illumination and radiation units*	261
4	*Atomic weights*	262
5	*Formulae and molecular weights of some commonly used chemicals*	262
6	*Chemical gravimetric conversions*	263
7	*Temperature conversions*	263
BIBLIOGRAPHY		264
INDEX		271

Chapter 1

Why change?

When plants are grown in containers their roots are restricted to a small volume of compost*, consequently the demands made on the compost for water, air and nutrients are much more intense than those made by border-grown plants which have an unrestricted root-run and an infinitely greater volume of soil in which to grow. It has long been recognised by amateur and commercial growers alike that simply using a garden or border soil in a pot without any improvement to its physical properties or nutrient status will give poor results. Growers have traditionally used such materials as leafmould, decayed animal manure, spent hops, peat, mortar rubble, wood ashes, sand and grit as additives to mineral soils to improve their physical properties. The introduction of the John Innes composts by Lawrence and Newell in the 1930s did much to standardise and rationalise the multitude of materials then being used. Forty years of experience with these composts have shown that, if prepared and used correctly, very good results can be obtained.

The compost is universally recognised as being one of the foundation stones upon which the successful growing of pot plants is built. Achieving the correct physical and nutritional conditions is just as important for crops such as the tomato, which only remain in the pots for a few weeks before being planted into the glasshouse borders, as it is for plants such as cyclamen which spend the whole of their lives in pots. Any mistake made during the seedling or young plant stage is not easily rectified; often a check in growth at this stage, caused by a soil sterilisation toxicity, salinity or nutritional problem, will still be evident when the plant is mature. Obviously, very careful consideration must therefore be given before making any funda-

---

* In horticultural terminology 'compost' has several meanings. It is used with respect to (a) plant material that has undergone biological decomposition before being dug into garden soil, (b) animal manure and straw which is used as the basis for mushroom growing and (c) mixtures of organic and mineral materials used for growing plants in containers. Either of the terms 'plant growing media' or 'plant substrate' would be less ambiguous and more precise for our purpose, but in view of its general acceptance and use by pot plant growers, the term 'compost' is used in that context in this book. In the USA the terms 'mix' or 'media' are used in place of compost.

mental changes to such an important part of pot plant culture as the compost; indeed, it might well be asked, 'Why change?'

There are essentially two reasons why pot plant growers have been changing from the loam based John Innes composts to composts made from peat, vermiculite, plastics, etc., and other loamless materials. They are:

(1) The practical difficulty of obtaining sufficient quantities of loam which conform to the John Innes specification.
(2) Economic advantages in dispensing with the mineral soil and its preparation.

To meet the requirements of the John Innes composts, the loam must be a 'medium clay', free from lime or chalk and preferably be a 'ley' or turf; this ensures that it has a good structure with sufficient organic matter or fibre to supply the plant with nitrogen over a long period. Such loams are difficult to obtain today. Legislation prevents the uncontrolled stripping and sale of agricultural soil, and soils for compost making are obtained from various sources, e.g. building sites, roadworks, etc. These soils are often quite unsuitable for compost making. They can be heavy waterlogged clays which are difficult to handle, light sandy soils with little or no structure, subsoils with no organic matter, or soils with a high chalk content. A further very important point to consider when selecting a soil for compost making is its reaction when steam-sterilised. To ensure that the loam is free from plant pathogens it is necessary that it is partially sterilised, preferably by heat treatment (Chapter 11), before it is used. Unfortunately, the beneficial effects of this treatment in eliminating the plant pathogens are accompanied by several biological and chemical changes, some of which can be detrimental to plants. One of the most important of these changes following sterilisation is the increase in the amount of water-soluble and exchangeable forms of manganese; in some soils this can reach toxic levels. Raising the pH of the soil by adding ground limestone or chalk, preferably before sterilisation, will materially reduce the amount of active manganese. Unfortunately this treatment may also have an accompanying adverse effect with respect to the forms of nitrogen present (Chapters 5 and 12). Steam or heat sterilisation of the soil also changes the balance of the various types of bacteria which convert the organic matter into mineral nitrogen. The ammonifying bacteria are spore forming types, they are resistant to heat treatment and multiply rapidly after the soil has been sterilised. The nitrifying bacteria, however, which convert the ammonium into nitrites and then into nitrates, are temporarily eliminated from the soil when it is heat treated; consequently one link in the normal process of nitrogen mineralisation is broken and a build-up of ammonium can occur. Steam sterlisation, a high soil pH and the presence of readily decomposable organic matter are the prime factors required to create free ammonia problems (see Chapters 5 and 11). The liming of a soil to reduce manganese toxicity following steam sterilisation is a good example of remedial action for one problem possibly creating another problem.

Before leaving the matter of loam selection there is also the question of continuity of supply to consider. A grower with one acre of glasshouses in a 'year-round' pot chrysanthemum programme, for example, will use about 4 cu. m (approximately 5 cu. yds) of compost each week, and this will require 120 cu. m of loam each year. To maintain a week-by-week consistency of plant quality calls for an equal consistency in the quality of the compost, and securing such a quantity of loam with a consistent quality obviously presents difficulties.

Regarding the economic aspects of loam and loamless composts, no critical comparisons can be made because of local differences in the basic prices of the materials and of differences in the degree of mechanisation available. One very large factor in the costs of preparing composts is that of heat sterilisation. Because various weed seeds and pathogens such as wireworms and damping-off fungi, etc. are normally present in mineral soils, it is essential that they are partially sterilised before use, preferably by steam. This means that a steam boiler and certain ancillary equipment is necessary (Chapter 11); such equipment is expensive both to purchase and to operate. Furthermore, the soil must be kept dry and under cover before it can be successfully treated. Not only does wet soil require considerably more steam than dry soil, it is much more difficult to treat efficiently because wet soils tend to 'puddle' and leave 'cold spots' of untreated soil. Diseases surviving in the untreated parts of the soil will then spread more rapidly into the treated soil and cause more trouble than if the soil had not been treated at all. By contrast, loamless composts based on peat, vermiculite or perlite do not usually require sterilisation, and peat packed in polythene-wrapped bales can be stored in the open and kept dry without the need for expensive storage facilities. Peat is not technically sterile, and it contains various microorganisms, but these are not usually pathogenic. Consequently in Britain and western Europe it is not customary to sterilise new peat before it is used. In the USA it is often considered advisable to steam the peat; much will depend on the source of the peat and its known history of pests and diseases.

Composts made from mineral soils are also much heavier to handle, both during their preparation and in filling and moving the pots or other containers. If large quantities of loam-based compost are required, it will be more desirable to use such mechanical equipment as soil shredders, conveyor belts, etc., in order to keep down the labour costs, than if the lighter loamless composts are used. For these reasons it is generally accepted that loamless composts are both easier and cheaper to prepare than conventional loam-based compost.

The advantages and disadvantages of loam-based and loamless composts can be summarized as follows:

1.1 LOAM COMPOSTS

ADVANTAGES

(1) The principal advantage is ease of plant nutrition. Composts made from a good loam have more plant nutrients and the nutrition, especially with respect to nitrogen and phosphorus, is easier.
(2) Minor element deficiencies are not common.

DISADVANTAGES

(1) Difficulty in obtaining suitable loam that does not give toxicities when steam sterilised.
(2) Continuity of supply and quality control.
(3) The loam must be stored dry and steam sterilised before it is used.
(4) The composts are heavy and difficult to handle.
(5) They are more expensive to prepare if done properly.

1.2 LOAMLESS COMPOSTS

ADVANTAGES

(1) A greater degree of standardisation of materials, less variability between successive batches of compost.
(2) Do not require steam sterilisation.
(3) Cheaper to prepare.
(4) Lighter to handle.
(5) The initial lower nutrient content of the materials can be used to give more controlled growth.

DISADVANTAGES

(1) Control of the supply of nitrogen, phosphorus and minor elements such as boron and copper is more critical.
(2) Greater dependence on liquid feeding.
(3) Lack of general 'buffer' capacity, i.e. they are more likely to show rapid changes in general nutrient levels.

The grower must now decide which of these two types of compost he wishes to use. A description of the materials available for making loamless compost, together with their preparation and fertiliser requirements, is given in Chapters 2 to 9, whilst the preparation of the more conventional loam-based John Innes compost is described in Chapters 11 and 12.

Chapter 2

Alternative materials

There are a number of bulky materials which can be used either separately or in various combinations to make loamless composts; the choice of a particular material is usually determined by its availability, cost and local experience of its use. In NW Europe peat is the material most widely used for compost making, whilst in the USA vermiculite and perlite are also extensively used.

The most important single factor when choosing a material for compost making is that it should be free from substances that are toxic to plants. If this requirement can be met, then a wide range of materials can be successfully used, providing that the management (and this usually means the watering and the nutrition) is adjusted to the requirements of the media and the crop. The common conclusion that '... this particular compost gave the best results' really means '... this compost gave the best results *under the particular system of management prevailing during the experiment*'. A change in management practice or even the seasonal change in the environment can often give completely different results.

The following is a brief description of the materials most commonly used for compost making.

2.1 PEAT

This is by far the most widely used material for making loamless composts; it can be used either by itself or in combination with other materials. In the raw state it is usually deficient in the principal plant nutrients, but in common with mineral soils it has the important advantage of having a high cation exchange capacity: this is a chemical mechanism which helps in regulating the supply of some nutrients to the plant (p. 60). From a casual inspection of peat in bales it might at first appear that it is a uniform and standardised product. Closer investigation however will soon show that this is not the case. Peat is formed by the partial decomposition of plants growing in areas having high rainfall, high humidity and low summer temperatures. Under acid, waterlogged conditions, and in the absence of nutrients, the micro-organisms, which would normally break down or decompose the plants, are excluded and only partial decomposition of the dead tissue occurs. Differences between peats are related to variations in local climate and in the species of plants from which the peats are formed.

TYPES OF PEAT

The most important peat-producing countries in western Europe are Germany, Finland, Sweden, Norway, Ireland and Scotland. Robertson (1971) has surveyed the nature and extent of the peat resources of Scotland, where peat covers approximately 11% of the land area, and Barry (1969) has reviewed the formation and distribution of peat in Ireland. Unfortunately there is no accepted international method of peat classification and terminology; each country has its own system and these have been reviewed by Farnham (1968). In Britain the following system of classification is used:

Low moor peat is formed from the shallow flooding of a depression or the infilling of a lake. In both cases the water contains some mineral bases. Deposits of this type may be formed at any altitude but they usually occur at fairly low elevations.

High moor peat is formed without prior inundation, the surface being kept continuously saturated because of the moisture-holding properties of the surface as well as heavy rainfall. It usually overlies peat of the low moor type. Essential conditions for its formation are comparatively cold, wet conditions and a very low level of mineral bases. Bogs of this type are generally higher at the centre than at the margins and are often referred to as raised bogs. They usually contain two layers: a lower one of more humified peat and an upper one of generally less decomposed material; both contain considerable proportions of sphagnum moss remains. In Germany these bogs are known as 'Hochmoor' and the peat can be 'white peat', i.e. slightly decomposed, or 'black peat', i.e. highly decomposed.

The term 'basin peat' can also be used to describe the low moor and high moor types and refers to the peats formed in and over the pool or lake from which the bog was formed.

Blanket bog is similar to high moor but is formed in areas where the surface is continuously saturated by high rainfall. It largely follows the contours of the ground and the level of mineral bases in the peat is very low. Blanket bogs generally occur on high moorland but they can also occur at sea level.

The raised bogs of the high moor type usually give the best peats for compost making purposes.

Several plants grow in the bogs from which the peat is formed and the peats are classified according to the principal plant species. For horticultural purposes they can be grouped under two headings: sphagnum and sedge peats.

Sphagnum. Several sphagnum mosses are found in peat bogs, the most important of which are *S. papillosum, S. imbricatum* and *S. magellanicum,* which are members of the *cymbifolia* group, and *S. rubellum, S. plumulosum* and *S. fuscum* which belong to the *acutifolia* group. Sphagnum leaves consist of a single layer of cells and those in the *cymbifolia* group have large boat-shaped leaves, this gives them a high rate of water absorption and retention. Peats from this group tend to be loose and bulky. Sphagnums in the *acutifolia* group have smaller leaves, retain less water and the resulting peats are denser. The general characteristics of the sphagnum peats are a spongy fibrous texture, a high porosity with high water retaining capacity, a low ash content and usually a low pH. Some sphagnum type peats contain variable amounts of cottongrass (*Eriophorum*).

Sedge peat. This type of peat is formed from sedges (*Carex* species) and reed grass (*Phragmites*) with some Hare's tail Cotton Grass (*Eriophorum vaginatum*) and common heather (*Calluna vulgaris*). Sedge peats have usually developed under the influence of drainage from mineral soils, consequently they contain more plant nutrients than the sphagnum peats. They are darker, more humified and decomposed than sphagnum peats and have a higher cation exchange capacity per unit weight. They have a lower water retaining capacity than sphagnum peats and whilst they make good agricultural land when reclaimed, they can also be used for making pot plant composts if given the correct management. Because of their structure and better binding capacity when compressed, peats of this type are preferred to the younger, less decomposed sphagnum types for making peat-blocks. These are free-standing units of fertilised, compressed peat used for propagating plants such as chrysanthemum and lettuce which are eventually planted into the glasshouse borders. The peat-block is in fact somewhat similar to the older type of soil block which was made from the John Innes composts.

In the USA the specification for peat introduced by the General Services Administration of the US Government in 1961 is being revised. The American Society for Testing Materials (ASTM) has proposed a new system of peat classification based on the generic origin and fibre content. There are five groups in the new system:

(1) *Sphagnum Moss Peat (Peat Moss).* An oven-dry peat sample contains over 75% by weight of Sphagnum moss fibre. The fibre should be stems and leaves of Sphagnum in which the fibrous and cellular structure is recognisable. The samples must contain a minimum of 90% of organic matter on a dry weight basis.

(2) *Hypnum Moss Peat.* The Hypnum moss fibre content must be more than 50% of the oven-dry weight of the sample, and the organic matter content

must be not less than 90% of the oven-dry weight. The fibres should be stems and leaves of various Hypnum mosses.

(3) *Reed-Sedge Peat*. An oven-dry sample of the peat must contain a minimum of 33·3% by weight of reed, sedge or grass fibres, i.e. non-moss fibres.

(4) *Peat Humus*. The total fibre content of the oven-dry peat contains less than 33·3% by weight of total fibre.

(5) *Other Peat*. Peats which are not classified by the previous groups.

For the purpose of this classification the term fibre is described as plant material retained on a 100 mesh sieve or larger. This includes stems, leaves or fragments of bog plants. Wood particles larger than 0·5 in. are excluded and also mineral matter such as stones, gravel, shells, etc.

In an attempt to introduce a system of peat classification that would be internationally acceptable, Kivinen and Puustjärvi (1972) have proposed the following system based on the botanical composition of the peat and also its degree of decomposition. There are five main groups in the system:

(1) Moss peat (Sphagnum moss peat).
(2) Hypnum moss peat.
(3) Sedge peat.
(4) Forest peat.
(5) Black peat (peat humus).

The three degrees of decomposition, based on the von Post scale, are:

Class	Degree of decomposition
Light peat	H 1–3
Dark peat	H 4–6
Black peat	H 7–10

COMPARISON OF PEAT TYPES

When the various types of peat have been made into composts by the addition of relatively large amounts of base fertilisers containing both macro and micro elements, any differences which formerly existed in the mineral levels of the peats will have been substantially reduced. Large differences will however still remain in the amount and type of organic compounds such as cellulose, lignins, proteins, etc., which are present in composts made from sphagnum and sedge type peats. This can have a marked effect upon the activity of certain plant growth regulators when they are either mixed into the composts or applied to the compost in solution. Certain plants such as chrysanthemum and poinsettia will normally be too tall to make acceptable pot plants unless they are treated chemically to prevent them attaining their full natural height. Growth regulators such as phosphonium chloride (Phosfon) and chlormequat chloride (Cycocel) are sometimes applied direct to the compost rather than used as foliar sprays, and the response of plants to these materials has been found to vary with the type of peat used.

The results of an experiment at the Glasshouse Crops Research Institute on the control of plant height by the addition of Phosfon at varying rates are shown in Figure 2.1. The composts used were made with 25% by volume of fine sand and 75% by volume of peat, the latter being either a sedge type of peat or sphagnum peats of Irish or Finnish origin. A mineral soil compost (John Innes) was also used as a control compost. Chrysanthemum variety 'Bright Golden Princess Anne' was grown for a mid-winter crop. When plants were grown in the sedge peat compost the growth regulator had very little effect upon the height of the plants, even when used at the maximum rate of 0·75 g/l of compost (1¼ lb./cu.yd). With the Irish peat compost there was a progressive decrease in the plant height as the growth regulator was increased from 0·25 g/l to 0·75 g/l, the optimal rate being about 0·5 g/l (¾ lb./cu.yd). With the Finnish peat compost, even at the lowest rate of application (0·25 g/l) the reduction in height was slightly greater than is desired commercially, whilst at the highest rate of application the plants would not be acceptable because of the excessive reduction in their height. In the John Innes compost, which contained approximately 60% by volume of a mineral soil, the effect was intermediate to that obtained with the sedge and Finnish peat composts.

A similar effect has been found when poinsettias were grown in composts made with different peats and then treated with chlormequat chloride (Cycocel) to control their height. When grown in a sedge peat compost and given a drench of 100 ml per pot of a Cycocel solution containing 2 500 ppm active ingredient, the height of the plants was reduced by 7·7 cm, whilst in a Finnish peat compost the same treatment reduced the height by 11 cm. At a strength of 5 000 ppm active ingredient the reduction in height was 10 and 14 cm respectively, i.e. with the sedge peat compost approximately double the strength of Cycocel was required to give the same effect as a single strength drench in the Finnish peat compost. Inactivation of certain herbicides is also known to occur when they are used on organic and peat soils.

Fig. 2.1 The effect of peat type on the response of chrysanthemum to the growth regulator 'Phosfon' mixed into the compost. *Top to bottom:* sedge, Irish and Finnish peat composts and John Innes compost. *Left to right:* control plants, Phosfon at 0·25, 0·50 and 0·75 g/l of compost.

24 Alternative Materials

Table 2.1. *Composition of three peats (percentage of total organic matter).*

Peat type	Hemi-celluloses	Cellulose	Lignin	Protein
Sphagnum/ Cotton grass	16·6	14·0	32·8	3·6
Scirpus type	22·4	13·8	41·4	4·1
Phragmite types	3·4	8·4	50·8	6·4

The difference between the young, less humified sphagnum peat and the sedge peat is believed to be related to their cellulose and lignin contents. Ogg (1939) found large differences in the compositions of three peats (Table 2.1).

PEAT DECOMPOSITION

The degree of peat decomposition can be measured in several ways. Two of the best known are the von Post and the volume weight methods. The von Post method consists of a scale of 10 grades designated H1 to H10. It is based on assessing the quality of the water which is exuded when wet peat is compressed in the hand, and also on the amount of peat which is pressed out between the fingers (Table 2.2). This test is sometimes supplemented by determining the volume weight or bulk density of the peat, i.e. the weight of peat per unit volume. A standard procedure is used to wet the peat and allow it to settle to a natural density without any compression. Puustjärvi (1970*b*) has shown a good correlation exists between the degree of

Table 2.2. *The von Post scale for measuring peat decomposition.*

Degree of decomposition	Quality of water exuded	Proportion of peat pressed out between fingers
1	Clear, colourless	None
2	Almost clear, yellowish brown	None
3	Turbid, brown	None
4	Very muddy brown	None
5	Extremely muddy, contains little peat in suspension	Very little
6	Dark, plenty of peat in suspension	One third
7	Very dark, thick	Half
8	Very thick	Two thirds
9	No free water	Nearly all
10	No free water	All

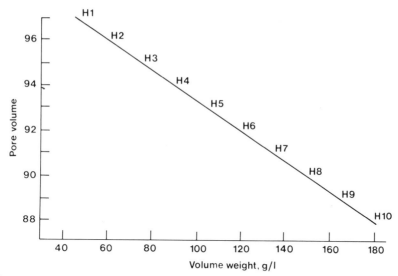

Fig. 2.2 Decomposition of peat. The relationship of the pore volume, the volume weight or bulk density and the von Post scale of decomposition. (From *Peat & Plant News*, 1970.)

decomposition as determined by the von Post scale and the volume weight method; this relationship is shown in Figure 2.2.

A general summary of the most important physical and chemical characteristics of sphagnum peat is given in Table 2.3.

For compost-making, peats having a medium granulation with particles ranging up to 9·5 mm ($\frac{3}{8}$ in.), 50% of which are between 1·6 mm ($\frac{1}{16}$ in.) and 6·3 mm ($\frac{1}{4}$ in.) diameter, are preferred.

Table 2.3. *Properties of sphagnum peat.*

Bulk density, g/l	60–100
Pore volume, %	> 96
Organic matter, %	> 98
Ash, %	< 2
Total nitrogen, % weight	0·5–2·5
Cation exchange capacity (me/100 g)	110–130
pH (in water)	3·5–4·0
Weight per bale	56 kg
	(125 lb.)
Volume per bale	360 l
	(12·5 cu. ft)
	(10 bushels)
Water content in bale,	
moist sample basis (%)	50–60
dry weight basis (%)	100–150

Alternative Materials

PEAT PRODUCTION

In recent years there have been considerable changes in the methods of peat production; it is now a highly mechanised process employing large and expensive equipment. A peat bog can occupy an area of several hundred acres and it is prepared for harvesting by first cutting a series of connecting drainage trenches. After being drained, the surface layer of moss and heather is removed and production can then commence. Two distinct methods are

Fig. 2.3 A triple drum peat milling machine. (Acknowledgement to Board Na Mona.)

used to harvest the peat from the bog: they are known as milling and sod cutting. The method chosen depends on the nature of the bog, whether it contains undecomposed tree trunks, etc. and the local climate. With the milling process, large rotary-cultivator type machines are used to loosen the top 15 mm ($\frac{1}{2}$ in.) layer of peat. This is allowed to dry before other machines form it into ridges and load it ready for transporting off the bog. This method of harvesting can only be used when the surface layer of the peat bog has dried out sufficiently to allow the peat to be milled; it is not a suitable method in areas where dry periods are short and infrequent.

With the sod system of harvesting, brick-shaped units of peat are cut from the vertical side of a trench by a machine which also lays the peat sods onto the surface of the bog to dry. After the initial drying, the sods are mechanically gathered to form loose hedges or ricks. With this system of harvesting, the peat is usually left on the bog to be subjected to a period of low temperature before it is taken to a factory for processing. Some examples of the equipment used in harvesting peat are shown in Figures 2.3 and 2.4.

A light railway is usually used to transport the peat from the bog to a factory for further processing, i.e. shredding (if necessary), sieving, grading and finally compression into polythene-wrapped bales. When most of the peat has been removed, the bog is usually brought into cultivation with vegetable or agricultural crops. It is then referred to as a 'cut-over' bog.

Fig. 2.4 Peat harvesting with a sod-cutting machine. (Acknowledgement to Board Na Mona.)

2.2 OTHER ORGANIC MATERIALS

Whilst peat is undoubtedly the most important of the organic materials used for compost making, several other organic materials are also used, such as pine needles, shredded bark, sawdust and the organic residues of crops, e.g. peanut hulls, rice hulls, coffee waste, etc. Often the popularity of a particular material will be related to its local availability, its cost and also the amount of local experience in its use. Certain materials are also traditionally used for particular crops, e.g. pine needles are extensively used by azalea growers in Germany and Belgium, whilst bark is frequently used by orchid growers. Success in using these materials in pot plant composts depends largely upon understanding their particular physical and chemical characteristics, both in formulating the composts and in their subsequent management.

Alternative Materials

SAWDUST AND SHREDDED BARK

Sawdust and bark waste have been used for mulching field crops for many years, and following the rapid expansion in the number of pot plants and hardy nursery stock plants now being grown in containers, growers have shown interest in these materials for compost making. Two problems are frequently encountered whenever these materials are used in any concentration in the composts; they are: (a) nitrogen deficiency and (b) the production of substances toxic to plants.

Table 2.4. *Average percentage of carbon released as carbon dioxide from woods and barks over sixty days.*

Type of wood	Wood carbon released as CO_2		Bark carbon released as CO_2	
	No nitrogen	Nitrogen added	No nitrogen	Nitrogen added
Softwood	12·8	12·0	8·8	8·2
Hardwood	30·3	45·1	22·4	24·5

The comparative value for wheat straw was 54·6%.

Both sawdust and bark contain large amounts of carbon and only small amounts of nitrogen, the average composition being approximately 50% of carbon and only 0·1% of nitrogen; bark usually contains slightly more nitrogen than sawdust. When decomposition starts, the bacteria are able to utilise the nitrogen more effectively than the plants, thereby creating a temporary nitrogen deficiency. Allison (1965) studied the rates at which these materials decompose in the soil by measuring the amount of carbon released as carbon dioxide over a sixty-day period; his results are summarised in Table 2.4. Whilst there was some variation between species in the rates of carbon dioxide production, softwoods had a significantly slower rate of decomposition than the hardwoods, the average rate for a number of softwood species being about one quarter of that of the hardwoods. The bark from softwoods also decomposed at a slower rate than hardwood bark and both groups of barks decomposed much more slowly than their respective woods.

The amount of nitrogen immobilised by the fungi and bacteria during the decomposition of these materials has been determined by Allison *et al.* (1963) (Table 2.5). They found that the average amount of nitrogen required during the decomposition of the softwoods was equivalent to 0·59% of the weight of the wood whilst for hardwoods the amount was 1·1%, i.e. approximately double. The corresponding value for wheat straw was 1·7%. Soft-

Table 2.5. *Rate of nitrogen immobilisation of various woods.*

	Nitrogen immobilised[1]			Decomposition rate	
Kind of wood	10 days	40 days	160 days	CO_2 produced[2]	Relative value
Redwood	0·13	0·21	0·34	5·3	100
Calif. incense cedar	0·17	0·52	0·52	4·9	92
Cypress	0·13	0·17	0·37	3·8	72
White fir	0·04	0·25	0·54	16·2	306
Douglas fir	0·07	0·07	0·30	11·2	211
Red cedar	0·17	0·17	0·41	3·9	74
Shortleaf pine	0·78	1·27	1·13	50·0	943
Western white pine	0·11	0·35	0·89	22·2	419
White pine	0·08	0·29	0·41	16·4	309
Ponderosa pine	0·05	0·19	0·42	13·7	258
Average of nineteen softwoods	0·14	0·33	0·59	14·3	
Average of nine hardwoods	0·78	1·21	1·10	45·1	

[1] Expressed as % of weight of wood.
[2] Expressed as % of added carbon.

woods were very much more variable in their nitrogen requirement than the hardwoods. For example, short leaf pine required more than three times as much nitrogen as redwood. With the hardwoods, however, there was not more than 15% difference between those species having the greatest and lowest nitrogen requirements. Not only will the rate of decomposition, and therefore the demand for nitrogen, depend upon the chemical composition of the material, i.e. the relative amount of lignin, cellulose and cork present, but it will also be controlled by the size of the particles. Finely ground bark will have a greater requirement for nitrogen than coarse particles of bark from the same species. With sawdust, the amount of nitrogen required will also depend upon the age of the sawdust and the amount of decomposition that has occurred naturally before the sawdust is used. For every 50 parts of carbon assimilated by the microbes that decompose sawdust and bark, there is also a requirement of 1 part of nitrogen, 0·5 parts of phosphorus and 0·1 part of sulphur. When untreated bark and sawdust is added to soils or is used with other materials to make potting composts, nitrogen is usually the limiting factor. When these materials are first composted in piles prior to being used for making potting composts, it is also necessary to add some phosphorus, usually in the form of superphosphate, and some sulphur as well as nitrogen.

In practice it is found that softwood sawdusts such as redwood and Douglas fir require about 1·8 kg of ammonium nitrate per cubic metre (3 lb./cu.yd), ponderosa pine requires 3 kg/cu.m (5 lb./cu.yd) and redwood bark 0·6 kg/cu.m (1 lb./cu.yd). Working with a mixture of hardwood barks, predominantly eastern cottonwood, oaks and silver maple, Gartner et al. (1973) recommended 3·6 kg of ammonium nitrate per cubic metre (6 lb./cu.yd).

The quantities of nitrogen required to offset that immobilised during the biological decomposition process are less than would have been predicted solely from the carbon:nitrogen ratios of the materials. This is because the woods and barks are relatively resistant to decomposition in comparison with straw, owing to the large amounts of lignocellulose they contain. Douglas fir sawdust with a C:N ratio of 620:1 decomposed slower and required less added nitrogen than wheat straw which had a C:N ratio of 373:1. Nevertheless the above rates of mineral nitrogen additions are much too high to allow the freshly treated sawdust or bark to be safely used in composts. At these rates of mineral nitrogen addition there would be the risk of salinity and toxicity problems with young plants. Whilst slow-release forms of nitrogen such as urea formaldehyde can be used at much higher rates than mineral nitrogen, it is normally safer to mix the nitrogen in mineral form with the moist sawdust or bark and then store the mixture for a period to allow decomposition to proceed before it is used for compost making. During this period, the mineral nitrogen will have been converted into organic nitrogen and the salinity will then have declined to a safe level.

Whilst there is some variation in the mineral composition of bark and sawdust samples obtained from different species of trees, in general the quantity of plant nutrients they contain is low in relation to that of a mineral soil. Comparison of the mineral contents of sawdust and bark from the Douglas fir, and of bark from a mixture of hardwoods (Table 2.6) shows that the bark of the Douglas fir contained significantly greater amounts of nitrogen, phosphorus and calcium than did sawdust from the same wood; the hardwood barks also had more nitrogen, phosphorus and magnesium

Table 2.6. *Average mineral composition of Douglas fir sawdust and bark and of a mixture of hardwood barks (% of dry weight).*

Material	N	P	K	Ca	Mg
Douglas fir sawdust[1]	0·04	0·006	0·09	0·12	0·01
Douglas fir bark	0·12	0·011	0·11	0·52	0·01
Hardwood bark mixture[2]	0·60	0·025	0·22	0·41	0·05
Sphagnum peat	2·50	0·030	0·04	0·20	0·15

[1] Bollen (1969).
[2] Cappaert et al. (1974).

than the Douglas fir bark. All the materials had significantly lower amounts of nitrogen and magnesium than a sample of sphagnum peat, but both barks had much higher calcium levels than the peat.

Most barks are acidic with pH values in the range 3·5 to 6·5. The quantity of lime added when they are composted is, however, relatively low. Normally no lime is added when composting hardwood barks as they contain relatively large amounts of calcium and the pH of the bark rises during the composting process. For softwood barks such as Douglas fir (*Pseudotsuga menziesii*), red fir (*Abies magnifica*) and white fir (*Abies concolor*) which are naturally more acidic and have a higher lime requirement, chalk is added at rates up to 6 lb./cu.yd (3·5 kg/cu.m). In their undecomposed state, sawdust and bark have a low cation exchange capacity, only about 8 milli equivalents per 100 g. As decomposition proceeds, this value rises to 60 or more milli equivalents per 100 g. Other changes which occur during decomposition of hardwood bark are an increase in the bulk density from $0·21\,g\,cm^{-3}$ to $0·26\,g\,cm^{-3}$, a slight reduction in the pore space and a significant increase in the volume of water available between tensions of 10 and 100 cm of water. There is also a reduction in the air content of the bark when measured at a tension of 10 cm.

Whilst the reduction in plant growth frequently associated with the use of sawdust and bark is often only the result of a deficiency of available nitrogen, certain woods are known to produce substances such as resins, turpentine and tannin which are toxic to plants. The wood of the Californian incense cedar (*Calocedrus decurrens*) is very toxic to young seedlings. Other woods which can cause toxicity are walnut, cedar and ponderosa pine. Mild toxicity can also occur with redwood shavings and sawdust: some cases of serious damage have been reported following steam sterilisation of this wood. The bark of the white pine is particularly toxic and amongst the hardwood barks, silver maple and black walnut have a high toxicity. The bark of the tuliptree (*Liriodendron tulipifera*), commonly known as yellow poplar in the USA, has also caused a slight toxicity. Gartner *et al.* (1973) have reported that bark removed from the logs in the middle of the winter season had a greater toxicity than bark collected at other times of the year. Ageing the bark in a moist condition for 30 days, during which time the heap was turned frequently, eliminated the toxicity.

In Britain the Forestry Commission has investigated the use of pulverised or ground softwood bark for potting composts (Aaron, 1974). It was found that the untreated bark of certain softwoods could be toxic to plants owing to the presence of volatile oils. Corsican pine and Scots pine had only a low toxicity, Japanese larch and European larch had an intermediate level of toxicity, whilst the barks of sitka spruce and Douglas fir showed the greatest toxicity. When the bark was pulverised and stockpiled, thermophyllic bacterial activity increased the temperature in the centre of the heap to about 60°C and this temperature was maintained for 6–8 weeks. After 12 weeks the volatile oil content of the bark had been reduced from 0·105% to 0·015% and the monoterpene content was then below the phytotoxic level. The shredding of the bark also fragments any rhizomorphs of the fungus *Armillaria* (honey fungus) which may be present and it is known that the rhizomorphs are incapable of surviving more than a few weeks in the fragmented form.

Sawdusts, barks and crop residues can all be successfully used in composts providing that adequate allowances are made for their higher nitrogen requirements, and possible toxic contents, by first composting the materials with the added nitrogen for 30 days before they are used.

COFUNA

Cofuna is a proprietary product produced in France and used in NW Europe. It is made from a mixture of oil cakes and seaweed which is ground, fermented and inoculated with a mixture of bacteria. These include the cellulolytic, ligninolytic and pectinolytic bacteria which break down fibrous material, and also the nitrogen-fixing bacteria which produce nitrates from the atmospheric nitrogen which plants are otherwise unable to utilise. Cofuna is used in some countries as an alternative to farmyard manure.

Its chemical analysis is:

Nitrogen (organic)	2·02%	Iron	0·14%
Nitrogen (ammoniacal)	0·15%	Manganese	0·002%
Phosphorus	0·23%	Copper	0·0003%
Potassium	0·51%	Zinc	0·01%
Sodium	0·13%	Chloride	0·10%
Calcium	1·94%	Molybdenum	9 ppm
Magnesium	0·13%	C/N ratio	22·7
	pH 7·20		

Number of micro-organisms per gram 2 700 000 000

Nelson (1972) compared the growth of chrysanthemums in composts containing Cofuna with those in other composts having various soil amendments. Tissue analysis of the leaves of plants grown in a compost having 33% by volume of Cofuna, but which did not receive any liquid fertilisers, showed that the levels of P, Ca, Mg and Fe were similar to those of plants grown in a compost without Cofuna, but which had been given weekly applications of liquid fertilisers. The nitrogen and potassium levels of the plants grown in the Cofuna compost were between 50 and 70% of those plants which had been given regular liquid feeding. This gives some indication of the amount of plant nutrients supplied by Cofuna. Cofuna should not be steam-sterilised and if it is necessary to sterilise the other ingredients of the compost this should be done before the Cofuna is added.

SEWAGE WASTE

To assist with the disposal of large quantities of sewage sludge or 'night soil', city and local authorities frequently make it available to farmers and growers at only a nominal cost. Whilst it has been successfully used in agriculture in controlled amounts on grassland and arable soils, its use in plot plant composts is not to be recommeneded because of the inherent risks of toxicity to plants. These sludges frequently contain large amounts of heavy metals and several instances of the loss of seedlings have been traced to the use of sludges in composts.

Nickel, zinc and chromium are often present in high concentrations in sludges and are known to be toxic to plants. Patterson (1971) surveyed the metal content of sewage sludge from 42 areas: the amounts of metals found by extraction in 0·5N acetic acid are given in Table 2.7.

Table 2.7. *The range of total and the 'extractable' amounts of nickel, zinc, chromium and copper found in sewage sludges.*

Element	Total, ppm	Extractable, ppm	Extractable, %
Nickel	20–5 000	6·8–320	14·7–92·7
Zinc	800–49 200	230–7 100	14·5–97·4
Chromium	40–8 800	0·9–170	0·7–8·5
Copper	200–8 000	2·9–460	0·5–30·9

Nickel is the most toxic of the heavy metals. Toxicities have been reported in soils having above 20 ppm of extractable nickel whereas in soils showing normal growth the extractable nickel level is about 5 ppm. Zinc toxicity is probably the most frequently occurring type following the use of sewage sludge. Comparable levels of extractable zinc (ppm) found in soils showing poor and good plant growth were:

	Poor growth	Good growth
Chrysanthemum	93	37
Wallflower	154	56
Bedding plants	150	100

Levels at which chromium becomes toxic to plants vary with the form of the chromium and the soil pH. It has been suggested that toxic effects can occur in soils when there is 10 ppm of chromium present as potassium dichromate, whilst with chromium trichloride 100 ppm of chromium are necessary to cause toxicity. Unlike most other metals, chromium is more toxic at high pH levels. Copper is not likely to cause toxicity below 150 ppm of Cu in the soil.

These high levels of heavy metals are usually associated with the effluents from industrial processes finding their way into the sewage sludge, e.g. chromium from electro-plating works and zinc from galvanising works and paint manufacture. High levels of zinc can also occur where newly galvanised water pipes are used. Toxic levels of some elements can exist in other materials such as dredgings from rivers and estuaries, and pulverised fuel ash.

2.3 MINERAL MATERIALS

SAND AND GRAVEL

Sand and gravel are seldom used by themselves as the basis of a compost; they are usually used in an admixture with peat for the purpose of changing the general physical properties, e.g. the bulk density and water retention. Providing they are free from clay and calcium carbonate, sand and gravel have no effect upon the chemical characteristics of the compost other than as dilutants. The difference between sand and gravel is purely one of particle size. For mechanical analysis purposes the definition adopted by the International Society of Soil Science is:

clay	below 0·002 mm
silt	0·002–0·02 mm
fine sand	0·02–0·2 mm
coarse sand	0·2–2·0 mm
gravel	above 2·0 mm

In horticultural practice the term fine sand is usually used for sands having particles in the range 0·05 to 0·5 mm diameter. A typical grade of fine sand used in a peat-sand compost would have a weight analysis of 100% passing through a 40 mesh BS sieve and 40% by weight passing through a 60-mesh sieve. A sand having fine particles blends with moist peat better than a coarse sand or gravel which tends to fall away from the root ball during repotting.

Whilst the sand or gravel by itself may have a good drainage rate and low water retention, it does not follow that it will have the same effect when used in a compost, for much will depend upon the general physical properties of the other constituents. The effects on the air and water relations of varying the peat-to-sand ratio and also the sizes of the peat and sand particles are described in the following chapter. Experimental work at the Glasshouse Crops Research Institute has shown that a wide range of sands can be used successfully *if the management of the crop is adjusted accordingly*. By far the most important point to observe in selecting a sand is its freedom from carbonates. These will cause a large rise in the pH of the compost and thereby create nutritional disorders, primarily affecting the availability of minor elements and the form and availability rate of organic nitrogen. Tomato seedlings germinated in a peat-sand compost made with a sand containing carbonates have shown blindness or the death of the growing point within 10 days of germination; these are typical boron deficiency symptoms. A simple test for the presence of carbonates is to add some dilute hydrochloric acid to the sand; the presence of carbonates is indicated by frothing and bubbling caused by the production of carbon dioxide.

Sand weighs about 1 600 kg/cu.m (100 lb./cu.ft) and one of its functions is to increase the bulk density of the compost. This can be an advantage in the case of tall plants, e.g. fuchsias, which have a tendency to topple over, especially when grown in light-weight plastic pots. In addition to giving density, the use of the correct grade of sand increases the absorbency of the compost; a compost made entirely from peat can be difficult to re-wet if it has been allowed to dry out too much.

CLAY

Powdered clay is sometimes used when making composts, the purpose being to increase the 'buffer capacity' or the resistance to sharp changes in nutrient levels. Clay has a high cation exchange capacity (see Chapter 4) but when peat, which also has a high cation exchange capacity, is used as the main component of the compost, the prime function of the clay is to regulate the supply of phosphorus and minor elements. Being in a powdered form rather than in its more natural aggregate form, the clay does not improve the physical properties of the compost. Indeed there is usually a reduction in the pore space and the rate of water drainage because of a decrease in the size of the pores if more than 10% by volume of powdered clay is used. As with the sand, it is very important to ensure that it does not contain any carbonates.

VERMICULITE

This is an aluminium–iron–magnesium silicate, which in its natural state is a thin plate-like or laminar mineral and resembles mica in appearance. Deposits are found in both the USA and South Africa and it is more widely used in these countries than in Europe. For horticultural purposes the mineral is first 'exfoliated', a term given to the process of heating previously graded particles to approximately 1 000°C, usually for about one minute. During this process the water trapped between the layers of mineral is rapidly converted to steam and the resulting increase in pressure causes the plates or layers to expand fifteen to twenty times their original volume, giving a lattice-like structure. In this form it has a high porosity value and a good air-to-water relationship.

The material is available in a number of grades, ranging from a fine particle grade for seed germination up to a grade with particles 6 mm ($\frac{1}{4}$ in.) in diameter; the average density is only about 80 kg/cu.m (5 lb./cu. ft). From the horticultural viewpoint vermiculite can be classified into two types; one is naturally slightly acid in reaction with a pH of about 6·0 to 6·8; the other type contains a significant amount of magnesium limestone which is broken into small pieces during exfoliation and thus raises the pH to above the neutral point. The former type of mineral is preferable for plant growing because of the usual difficulties with mineral nutrition which occur in alkaline conditions. Whilst the alkaline type of material can be used for compost making if it is first treated with phosphoric acid or monoammonium phosphate to reduce its pH, samples having a high magnesium limestone content and consequently a high pH are best kept for industrial purposes such as the thermal insulation of roofs, etc. Vermiculite has a relatively high cation exchange capacity, about 100 to 150 me/100 g and compares favourably with peat in this respect. Most samples contain 5–8% of available potassium and 9–12% of magnesium; composts containing vermiculite therefore require less of these minerals in the base fertiliser.

Vermiculite does not adsorb the anions Cl^-, NO_3^- and SO_4^{--} but it does adsorb some PO_4^{---}. Bylov *et al* (1971) have reported that when vermiculite was treated with a solution of potassium dihydrogen phosphate, between 63 and 77% of the phosphorus was adsorbed, the actual amount depending upon the type of vermiculite; 25% of the adsorbed phosphorus was retained in an available form and 75% was in a fixed or unavailable form. The phosphorus formed insoluble compounds with sesquioxides and magnesium. Vermiculite is also able to fix large quantities of ammonium in an unavailable form; this helps in regulating the amount of nitrogen available to plants when large amounts of organic or ammonium-producing fertilisers are used. Most of the 'fixed' ammonium is, however, available to bacteria and is converted to nitrate within a few weeks. It is then available to plants.

When vermiculite is used by itself as a growing medium for long-term cropping, there is a tendency for the lattices or honeycomb structure to collapse, resulting in reduced aeration and drainage, thus rendering the material 'soggy'. For this reason it is advisable to mix either some perlite or peat with the vermiculite as is done with the Cornell 'Peatlite' mixes or composts (Boodley and Sheldrake, 1972).

PERLITE

Perlite is an alumino-silicate of volcanic origin and is widely used in the USA and New Zealand, both countries having large natural deposits of this mineral. When crushed and heated rapidly to 1 000°C it expands to form white, lightweight aggregates with a closed cellular structure; these aggregates are stable and do not break down in the compost. The average density of perlite is 128 kg/cu.m (8 lb./cu.ft) and is available in a range of graded particle sizes. Because of its closed cellular structure, water is retained only on the surface of the aggregates or in the pore spaces between the aggregates. This means that composts with a high proportion of perlite are usually well drained and do not retain much water. Warren Wilson and Tunny (1965) found that the volume of water present between container capacity and the wilting point was 7% for perlite and 42% for vermiculite; when plants were grown in perlite, a capillary watering system gave better results than frequent hand watering. White and Mastalerz (1966) found water availability values of 34% for perlite and 59% for peat; the available water content of perlite–peat mixtures increased as the perlite content was decreased and the peat correspondingly increased. The actual amount of water retained by any medium is of course largely dependent upon the pore size distribution and the actual water availability in perlite will depend upon the grade of material used. It is general experience however that composts containing perlite are well aerated and have a lower available water content. For this reason perlite is often mixed with peat and used as a rooting medium for carnation cuttings, because its open structure prevents the occurrence of water logging from the frequent mist application of water to the cuttings during rooting. The low bulk density of the mixture also means that the delicate roots of the carnation are not so easily broken off during handling.

The chemical characteristics of perlite can be summarised by saying that it has virtually no cation exchange capacity; Morrison *et al.* (1960) reported a value of only 1·5 milli equivalents per 100 g. It is composed mostly of silicon dioxide (73%) and aluminium oxide (13%), and for practical purposes it can be considered devoid of plant nutrients. Plants grown in composts containing large amounts of perlite are, therefore, largely dependent upon liquid feeding. Green (1968) found that carnations grown in perlite suffered from aluminium toxicity when the pH of the nutrient solution was below 5·0; above this value no toxicity was observed.

STONE OR ROCK WOOL

In Denmark a stone wool fibre is used to make blocks or cubes for rooting cuttings and growing young plants. This material, which is largely an aluminium silicate with some calcium and magnesium also present, is smelted to approximately 1 500°C and fibres are formed from the molten material as it is cooled. In its prepared state it has a pore volume of about 97% and its function is to provide root anchorage for the plant and to regulate the water and air supply. It does not contain any plant nutrients and the plant must rely entirely on the inclusion of nutrients in the water supply. This system of growing is in fact a modification of the sand or gravel culture system.

PUMICE

Pumice is an alumina silicate of volcanic origin, it contains some potassium and sodium but only traces of calcium, magnesium and iron. The material is very porous, the pores being formed by the escape of steam or gas when the lava is cooling. It is sometimes used as a physical conditioner in potting composts or as an alternative to sand or gravel in hydroponic cultures. The particles are not very stable however and break down easily. Whilst in its natural state the material contains only small amounts of plant nutrients. It is, however, able to adsorb some calcium, potassium, magnesium and phosphorus from the soil solution, and release them to the plant later.

2.4 PLASTICS

One result of the rapid growth of the plastics industry has been the increasing interest shown in the use of foam plastics in pot plant composts. Initially waste polystyrene from industrial processes was used to make plastic flower pots and within a few years this type of pot had virtually displaced the traditional porous clay pot (Chapter 14). More recently, growers have been experimenting with composts made from peat mixed with plastic materials in foam or expanded form.

EXPANDED POLYSTYRENE FLAKES

Flakes of expanded polystyrene 4 to 12 mm in diameter were first used in agriculture for draining and improving the physical condition of heavy clay soils. The material is chemically neutral, it does not decompose or compress in normal use and has a low density, weighing only 20 kg/cu.m (34 lb./cu. yd). Flakes are formed from a large number of small closed cells and although the total porosity of the material may be as high as 95%, water is not absorbed within the flake. Thus, when used in potting composts it has the effect of reducing the amount of water retained and thereby it improves the compost aeration. Usually, the polystyrene is mixed with peat, the actual ratio of polystyrene to peat being adjusted to the type of plant grown. For example, composts for cyclamen, gloxinia, fuchsia, etc., usually contain 25% by volume of polystyrene and 75% of peat, whilst for epiphytic plants which require less water, the polystyrene content is increased to 50%.

It is important to remember that this material does not contain any plant nutrients and neither will it absorb or retain any from the fertilisers. Some allowance must therefore be made by starting liquid feeding earlier than usual. Composts containing expanded polystyrene are not so liable to overwatering; this can be an advantage when automatic watering systems are used. Composts made with this material must not be steam sterilised and neither should chemical sterilisers such as chloropicrin and methyl bromide be used.

This material is sold in Germany under the trade name 'Styromull' and at present it is more widely used by pot plant growers in Germany and Belgium than in the United Kingdom.

UREA FORMALDEHYDE FOAM RESINS

From the horticultural viewpoint the principal difference between the expanded polystyrene and the urea formaldehyde foams is the ability of the latter material to absorb water. It has an open cellular structure and can absorb between 50 and 70% of its volume of water. It has a low density, weighing 10–30 kg/cu.m (16–50 lb./cu.yd) and is available in both flake and block form. The formaldehyde content is less than 2·5%, but it is important to ensure that any free formaldehyde which may be present in the recently manufactured material is allowed to evaporate before the foam is used in making potting composts.

Urea formaldehyde foams do not have such high stability in soils as expanded polystyrene. Under acid conditions the annual rate of degradation is 15 to 20%, and the 30% by weight of nitrogen in the foam then becomes available to the plants during decomposition; there is also a very small amount of free nitrogen (0·25%) present in the material after manufacture. Because of the foam's very low density and slow rate of decomposition, the rate at which nitrogen is made available to the plants is, however, not very high. The material has negligible amounts of other nutrients and the pH is about 3·0. Thus whilst the physical properties of urea formaldehyde are somewhat similar to peat, the chemical or nutritional characteristics of the two materials are quite dissimilar. It is normally used in composts at 20 to 50% by volume and, as with the expanded polystyrene, careful attention to plant nutrition is required.

This material is sold in Germany under the trade name 'Hygromull' and as 'Floramull' in the USA.

POLYURETHANE FOAM

The most recent of the plastic foams to be used in plot plant composts is polyurethane. In common with other foams, it has a low bulk density, weighing 12–15 kg/cu.m (20–25 lb./cu.yd). It has an open cellular structure with a maximum water-retaining capacity of 70% of its volume. The pH of the foam is approximately neutral, and it is not decomposed by microorganisms, nor does it contain any plant nutrients. It is available in the usual flake form and also as cubes or blocks with a shallow depression in the upper surface into which seedlings or cuttings can be inserted. The blocks or cubes are usually partially joined together at the base to form a sheet and are easily separated when the plants require spacing; the blocks readily absorb water, either by capillary absorption or by overhead spraying. As the material does not contain any nutrients, all the plants' requirements must be supplied by liquid feeding.

A polyurethane foam in block form and known as 'Baystrat' is made by Bayer Chemicals in Germany; in the USA the Dow Chemical Co. introduced a polyurethane foam containing plant nutrients and known as Nutri-Foam.

42 *Alternative Materials*

The effects of the various foam plastics in pot plant composts on the air–water relationships have been reported by De Boodt and Verdonck (1971).

ION EXCHANGE RESINS

Water-soluble fertilisers create two types of problem in loamless composts. They can result in high salinity or osmotic values or they can be easily removed by leaching from many types of composts (p. 65). These problems have been approached experimentally by the use of resins with ion exchange properties in the composts. These resins retain cations such as potassium, ammonium, calcium, etc., and also anions such as nitrate, phosphate, etc., in an exchangeable form.

Nutrient-impregnated resins have been used in composts at rates from 2 to 10% by volume, the rate being adjusted to the anticipated demand for nutrients by the plant and the length of the growing period. Nutrients are released from the resin by an exchange with other ions already present in the irrigation water. To be effective, the irrigation water must have sufficient salts present to give a salinity of between 200 and 1 200 micromhos (0·2 to 1·2 millimhos); if the salinity falls below 200 micromhos, the rate of ion release from the resin may not be sufficient to meet the plant's requirements. At present there is insufficient experimental data upon which to base recommendations for the general use of nutrient-impregnated resins in pot plant composts.

Typical dry weight values of some newer materials used in composts are given in Table 2.8.

Table 2.8. *Dry weights of some compost making materials.*

Material	Dry weight lb/cu.ft	Dry weight kg/cu.m	Porosity
Bark, fir ($< \frac{1}{8}$ in)	14	224	69
Bark, redwood ($< \frac{3}{8}$ in)	8	128	80
Peat, sphagnum	6·5	104	90
Peat, sedge	14	224	78
Perlite	6–8	96–128	75
Plastic foams	0·5–2	8–32	Variable
Pumice	30	480	65
Rice hulls	6·5	104	80
Sand	100	1 600	40
Sawdust	12	192	78
Vermiculite	5–7	80–112	80

Chapter 3

Physical aspects

From the physical viewpoint, the compost can be regarded as comprising solid matter interspaced with voids or pores. Its function is to provide mechanical support for the plant and, at the same time, to make water and air available to the roots. The solid matter may be organic or mineral in origin and can occupy from 5 to 40% of the total volume. Mineral soils usually have a low pore volume, whilst potting composts with a high organic matter content may have a pore volume as high as 95%. The total pore volume of a particular sample of compost cannot, however, be regarded as an intrinsic value for that compost. An increase in the bulk density by firmer potting will materially reduce the total pore space and also change the air–water relationship of the compost.

3.1 PHYSICAL TERMINOLOGY

Before examining the compost from the physical viewpoint it is necessary to define some of the more important physical terms used:

air capacity – The proportion of the volume of compost that contains air after it has been saturated with water and allowed to drain.
available water content – The amount of water present after the compost has been saturated and allowed to drain, less the amount still present at the permanent wilting percentage.
bulk density – The dry mass per unit volume of moist compost.
capillary water – Water that is retained in the small pores by surface tension and moves as a result of capillary forces.
container capacity – The amount of water present after the compost in a container has been saturated and allowed to drain. The amount of water retained will be greater than that present in an agricultural soil at 'field capacity'.
desorption curve – See soil water tension curve.
easily available water – The amount of water held by a compost after it has been saturated with water and allowed to drain, minus the amount of

water present at some defined water tension. This is often taken as the volume of water between 10 and 100 cm tensions.

matric tension – See soil water tension.

osmotic tension – The decrease in the free energy of the soil water caused by the presence of dissolved salts. It is the negative pressure to which water must be subjected in order to be at equilibrium with the soil solution through a semipermeable membrane.

permanent wilting percentage – The amount of water still present in the compost when wilting plants are unable to extract sufficient water to regain turgidity even when placed in an atmosphere saturated with water vapour.

pF – The logarithm of the soil water tension expressed in centimetres height of a column of water, e.g. 100 cm water tension = pF 2, and the permanent wilting percentage = 15 450 cm water tension = pF 4·2.

pore space – The total volume of compost not occupied by mineral or organic particles.

soil water tension – The energy or negative pressure with which water is retained in the soil. It is the force per unit area that must be applied to remove water from the soil at any given water content. It does not include the water osmotic tension or salt effect.

soil water tension curve – A graph showing how the amount of water retained by a soil varies with tension or applied suction pressure. The units used are those of pressure; the low tensions present in wet soils are usually measured in terms of the equivalent heights of columns of water or mercury, whilst the higher tensions in drier soils are expressed in atmospheres.

solute stress – See osmotic tension.

specific gravity – The ratio of the bulk density or volume weight of the compost to the mass of unit volume of water.

total soil moisture stress – The sum of the soil water tension and the osmotic tension. In soils with very low nutrient levels there will be little difference between the soil water tension and the total soil moisture stress. In horticultural soils with high nutrient levels, and especially with loamless composts, the osmotic or solute tension will often exceed the soil water tension or matric tension and thus form the most important part of the total soil moisture stress.

volume weight – See bulk density.

water capacity – The difference between the quantity of water present after the compost has been saturated and allowed to drain, and the amount present at the permanent wilting percentage.

In addition to their obvious purpose for providing the basis for a physical description of compost, some of these parameters are also important for chemical or nutritional reasons. For example, the bulk density of a compost is an important factor to consider in interpreting the results of a chemical analysis if this has been made on the traditional weight basis and the results expressed as parts per million by weight. A mineral soil compost has a bulk density of about 1 g cm^{-3}, a peat–sand compost with 75% by volume of peat and 25% of sand has a bulk density of 0.45 g cm^{-3} and an all peat compost has a bulk density of about 0.1 g cm^{-3}. If the conventional practice has been followed of adding the fertilisers on a *volume basis* and the chemical analysis made on a *weight basis*, the peat compost will apparently contain much more fertiliser than the mineral soil compost. Either a correction factor must be applied for the different bulk densities of the composts or the analysis must be made on a volume basis. The lower the bulk density, the more important this becomes.

3.2 PHYSICAL REQUIREMENTS OF COMPOSTS

The physical requirements of pot plant composts are very different from those of agricultural soils. In agriculture, field-grown crops have a larger volume of soil from which moisture can be obtained either by root extension or by the capillary flow of water into the root zone. In pot plant culture, however, there is a relatively small volume of compost from which the plant's high demands for water must be met. This small volume of compost is soon filled with roots and there is then no further room for root extension. Under these conditions, any capillary flow of water within the compost is not of great importance. Experiments conducted at the Glasshouse Crops Research Institute to determine the total water and nutrient requirement of pot-grown chrysanthemums have shown that, for the normal commercial type of plant grown from five rooted cuttings in a 14 cm ($5\frac{1}{2}$ in.) pot holding one litre of compost, the average daily water requirement of a mature plant in summer can exceed 300 ml, i.e. equivalent to one third or more of the volume of the pot. This is the actual quantity of water lost by evapotranspiration and does not make any allowance for loss by drainage. In order to maintain a high growth rate it is necessary to keep a low water stress or tension in the compost. De Boodt and Verdonk (1972) have suggested that for optimal results the water stress in the compost should not exceed 100 cm: this is approximately equivalent to 7 cm of mercury or 10 centibars.

Whilst it is desirable for the compost to hold a large quantity of water, this must not be at the expense of adequate aeration. Roots require oxygen for respiration in the same way as leaves and other organs. After irrigation, the compost must have some free air space, otherwise the roots will be damaged by lack of oxygen and growth will be reduced. In soils that have been in a waterlogged or anaerobic condition for 2–3 days, ethylene may be present at 10 ppm and this is sufficient to completely inhibit growth. In such conditions, plants lose turgidity and wilt even when the compost contains sufficient moisture. Also, in waterlogged soils manganese is reduced to the divalent manganous form and toxic levels of available manganese can soon develop. Good aeration is therefore necessary to enable a sufficiently high rate of gaseous diffusion to replace the oxygen used up by the plant roots and the bacteria, and to prevent the carbon dioxide produced from reaching a toxic concentration. Plants such as Primula sinensis and antirrhinum are particularly susceptible to overwatering during winter, especially when grown in non-porous plastic pots (Chapter 14) and in composts with a low air capacity. The objective, therefore, must be to provide a compost which has both a high total porosity with a good degree of water retention and an adequate air capacity.

3.3 ENERGY CONCEPT OF WATER IN COMPOSTS

Because of the large variation in the bulk density, the total pore space and the pore size distribution, expression of the amount of water present in soils on a weight basis can be either meaningless or even misleading. For example, a heavy clay soil may contain more water than a light sandy soil and yet have less water which is available to plants. To overcome this difficulty, the water status of soils is often expressed in terms of water availability or total soil moisture stress. The total soil moisture stress (TSMS) comprises the matric tension or soil water tension and the solute stress or osmotic tension:

$$\text{TSMS} = \text{matric tension} + \text{solute stress}$$

Both the matric tension and solute stress are very important in pot plant cultivation, but for the present we will consider only the matric tension.

MATRIC TENSION

The pore volume of a compost contains air and water which change in inverse proportions. After irrigation, the compost will be practically filled with water, and the water or matric tension will be very low, say, equivalent to 1 cm of water tension or 0·001 of an atmosphere. In this condition, water is easily removed from the compost. Some will be lost by drainage, and when this has ceased, the water content will be at 'container capacity'. The water tension will then have increased slightly to, say, 10 cm of water or 0·01 of an atmosphere. For reasons to be given later, this condition is wetter than the analogous condition of 'field capacity' of agricultural soils. As more water is removed from the compost by evaporation and transpiration, the large pores are the first to give up their water, which is replaced by air. The pores do not, however, drain completely; a film of water is left which surrounds each soil particle and this water film decreases in thickness as the compost dries out. The small pores will be the last to release their water. Eventually, the water content of the compost will be so low, that even when wilting plants are placed in an atmosphere saturated with water vapour, they continue to wilt and do not regain their turgidity. This is known as the 'permanent wilting percentage'.

The force with which water is retained by a compost can be measured by means of a tensiometer (Fig. 3.1). This consists of a porous cup filled with water and connected to either a vacuum gauge or a mercury manometer. The cup is placed in the compost and the water in the tensiometer comes into equilibrium with the water in the soil. As water is removed from the soil, water leaves the tensiometer and the vacuum developed is registered on the vacuum gauge.

A modified form of this equipment can be used in the laboratory to construct a soil water tension curve. A typical water tension curve for a compost is shown in Figure 3.2. It will be seen that the relationship is logarithmic.

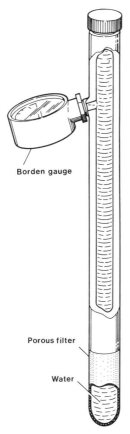

Fig. 3.1 Tensiometers can be used to measure the water tension in the compost. They are usually calibrated in centimetres of mercury or in centibars.

48 *Physical Aspects*

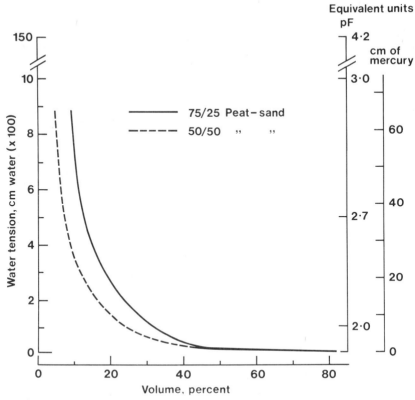

Fig. 3.2 A water tension curve of two composts, showing the relationship of the amount of available water with the tension.

Usually 75% or more of the available water will be released below a tension of one atmosphere. To continue the soil moisture tension curve beyond this tension requires different equipment. Usually a pressure membrane cell operated by compressed air is used to reach tensions up to 15 atmospheres.

The matric tension can be expressed in various units of pressure, the most commonly used units being centimetres of water; to avoid working in very large values when dealing with dry soils, the common logarithm of the column of water is often used and this is referred to as the pF. Other units which are sometimes used in measuring soil moisture stress, together with their equivalent values, are given in Table 3.1 and the approximate relationship between the scales is given in Table 3.2. For practical purposes, the bar and the atmosphere can be regarded as being equivalent to a column of water 1 000 cm in height. We have already seen that in pot plant culture the grower is only really concerned with low soil water tensions. Generally, he will not be

Table 3.1. *Equivalent units used to measure soil moisture tension.*

Bars	Ergs g^{-1}	Atmospheres	Cm of water
1·0	1·0 × 10^6	0·987	1·017 × 10^3
1·013	1·013 × 10^6	1·0	1·030 × 10^3
9·833 × 10^{-4}	9·833 × 10^2	9·703 × 10^{-4}	1·0
1·0 × 10^{-6}	1·0	0·987 × 10^{-6}	1·017 × 10^{-3}

interested in tensions above 1 000 cm (1 atmosphere) and will often be growing at tensions between 10 and 100 cm. High water tension will only be employed for special purposes, e.g. to harden off bedding plants before they are removed from the glasshouse, or to provide a check to vegetative growth and thereby assist in the setting of tomato fruit under poor light conditions in winter.

Table 3.2. *Approximate relationship between scales commonly used to measure soil moisture tension.*

Cm of water	pF	Cm of mercury	Atmosphere or bar
1	0	—	0·001
10	1	0·74	0·01
100	2	7·4	0·1
1 000	3	74	1
15 000	4·2	—	15

1 atmosphere is approximately equivalent to 1 bar or to 1 000 cm of water.

SOLUTE STRESS

Salts dissolved in the soil solution decrease the availability of the water to plants. The measurement of solute or salt stress and its interpretation and effect on plant growth will be discussed more fully in Chapter 4. For the present only the interaction of the matrix and solute stress on the total soil moisture stress will be discussed.

The solute stress or salinity of a solution is not only dependent upon the amounts of salts that are present, it is also directly related to the amount of water. A reduction of 50% in the amount of water will approximately double the salt concentration. Soils and composts contain some ions such as calcium and sulphate which have a low solubility, and a reduction of 50% in the amount of water will not necessarily double their concentration. Other ions in the soil solution such as potassium and nitrate respond in a more direct manner to a reduction in the water content.

This general relationship of the solute stress to the amount of water is very important for the practical pot plant grower. The shape of the soil moisture tension curve in Figure 3.2 shows that, when the compost is wet and has a low matric tension, the moisture content can be reduced by 50% with only a very small increase in the matric tension. Sometimes the increase in the matric tension will be only about 10 cm. Such a small increase will have very little *direct* effect upon plant growth. If the salinity of the compost is high by comparison with border and field soils, and this is usually the case, then the *indirect* effect of halving the moisture content is to approximately double the salt stress, which will have a much greater effect on plant growth than the small increase in matric stress would at first indicate. *When the salt stress in the compost is high, usually as a result of liquid feeding, it is very important to ensure that the compost does not dry out.* The importance of matric and solute stress and their integrated effect on plant growth is discussed more fully in Chapter 4. It is important to remember that tensiometers will measure only matric suction and do not measure either the solute or total soil moisture stress.

WATER ABSORPTION BY PLANTS

Whilst it is convenient to use the energy concept in examining the factors contributing to the water stress in soils, it is not generally considered that the expenditure of energy by plants plays an important role in enabling them to obtain their water from the compost. Several mechanisms are believed to operate in the absorption of water by plants and these fall into two groups, i.e. *passive* and *active* absorption. Passive absorption occurs when the transpirational loss by the leaves sets up a water tension gradient in the stem and roots. Water can be absorbed by this means even when the root system is dead or the roots have been removed. Active absorption includes water absorbed by osmosis, i.e. water moving from a low osmotic tension in the compost into the cells of the young roots which have a higher osmotic tension. Passive absorption is, however, generally regarded as being of greater importance than active absorption.

3.4 WATER ABSORPTION AND RELEASE BY COMPOSTS

The different manner in which water is absorbed and released by composts is not always appreciated by their users. When water is applied to a dry compost, a relatively shallow surface layer is first brought from the dry state to almost saturation point. If insufficient water is added to wet all of the compost in the pot, there will then be two layers having different water contents, a wet layer on top and a dry layer underneath. If more water is added to the pot, the wet front will move down through the compost until all the compost is wet and any additional water will be lost to drainage. The only time when wetting of the compost in the pot does not occur in this manner is when a capillary system of irrigation is used (Chapter 10). Therefore, when water is added to dry compost, there is no gradual and uniform increase in the water content. This can be achieved only under laboratory conditions by adding a small amount of water to the compost which is then thoroughly mixed and stored in a sealed container for a period, to allow the different moisture levels to come into equilibrium.

However, when plant roots remove water from the compost, it dries out at a relatively uniform rate. Differences in the moisture content of the compost within the pot may occur because of a number of factors, such as evaporation from the compost surface or pot wall, or to differences in the root concentration within the compost. The resulting gradient in the water content of the compost will, however, be relatively small. For practical purposes, it may be said that the compost will dry out at a relatively even rate, with a progressive change from the wet to the dry state similar to the desorption curve shown in Figure 3.2.

WETTING OF PEAT

One problem familiar to all users of peat is the difficulty in wetting it once it has dried out. This applies equally to newly opened air dry bales of peat and to peat-based composts which have been allowed to dry out in the pots. Sedge peats can be especially difficult to rewet.

Peat can be wetted by first breaking up the compressed bale, loosening the peat and then spreading it under a water sprinkler; turning or mixing the peat a couple of times whilst it is wetting will hasten the water absorption process. When peat or peat–sand composts are to be prepared, however, great care must be taken not to overwet the peat. If it is too wet, the base fertilisers will tend to adhere to the first particles of wet peat they touch and then remain in localised high concentrations rather than spreading uniformly through the compost during the mixing. This potential lack of uniformity in the nutrient content of peat-based composts can give rise to acute variation in the growth of plants in different pots. Initially, the peat has a low nutrient content, and poor mixing of the fertilisers can lead to the starvation of some plants, whilst the localised concentrations of highly soluble fertilisers can lead to toxicities in other instances. The moisture content of peat to be used in preparing loamless composts should be about 250% of the oven dry weight. The most convenient method of wetting the peat, when rotary mixers are used to mix the compost, is to first mix the dry peat with the sand and fertilisers, and then add a controlled amount of water whilst the mixer is still operating.

Neither of these methods can be used to re-wet the compost once it is in the pot, and one way of improving the rate at which the compost will absorb water, once it has been allowed to dry out, is to include a wetting agent or surfactant during the mixing. The difficulty in wetting peat is believed to be primarily due to a film of air strongly adsorbed on the surface of the peat and to the iron humates which are present, rather than to the natural waxes and resins which are also present in the peat. The surfactants operate by lowering the surface tensions of the water. Unfortunately some of these materials are toxic to young seedlings and great care must be taken to check on their toxicity before they are used. Sheldrake and Matkin (1971) screened a number of materials and three compounds which showed good wetting efficiency with a low risk of phytotoxicity were:

(1) T-Det C-30*[1], based on castor oil and ethylene oxide.
(2) Triton N-101[2], based on nonylphenoxy polyethoxy ethanol with 9–10 moles of ethylene oxide.
(3) Triton X-102[2], based on octyl phenoxy polyethoxy ethanol with 12–13 moles of ethylene oxide.

In the United Kingdom a material known as:

NONIDET LE[3], a primary alcohol ethylene oxide derivative,

has been successfully used.

These compounds are usually used at the rate of 0·1% or 1 000 ppm (1 fluid ounce per 6 gallons). Composts made from a mixture of peat and sand are not as difficult to re-wet as composts made only from peat.

EFFECT OF CONTAINER DEPTH ON WATER RETENTION

The air capacity can be a very useful parameter in making a physical assessment of a compost, especially its susceptibility to overwatering or remaining wet after irrigation. Before considering those factors which control the general air–water relationships and the air capacity of composts, it is important to realise that placing a compost in a shallow container, such as a pot, materially changes its air–water relationship by comparison with the same compost used in a glasshouse border where free drainage is unrestricted. After soil in the garden or glasshouse border has been saturated either by rain or irrigation, the water held in the large pores in the surface layers is removed by gravity and drains down into the sub-soil; this process is relatively rapid at first and then continues at a slow rate for a long period. In a pot, however, this natural drainage process is restricted by the presence of the base of the pot. Even though drainage holes have been provided, they will release only the free water which has accumulated in the base of the pot; they do not permit the natural drainage or flow of water which occurs in a deep profile. For this reason, composts in containers will contain more water after drainage has ceased than border soils, and the term 'container capacity' denotes a moister condition than 'field capacity'.

* These materials are available from:
[1] Thompson-Hayward Chemical Co., 5200 Speaker Road, Kansas City, Kansas, USA.
[2] Rohm and Haas Co., Independence Hall, Philadelphia, Pennsylvannia 19105, USA.
[3] Shell Chemicals Ltd, Villiers House, 41–47 Strand, London WC2N 5LA.

The effect of the depth of the container on the amount of air present, after a compost has been watered and allowed to drain, is shown in Figure 3.3. Increasing the depth of the container from 5 cm, which is equivalent to a 2 in. seed box, to 20 cm, which is equivalent to a pot 8 in. deep, increased the air capacity from 4 to 8% (Bunt, 1961). The actual relationship of air capacity

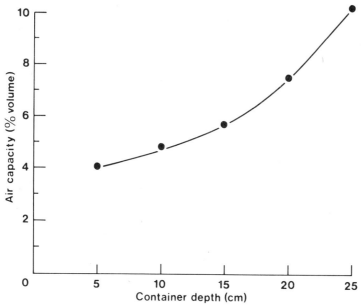

Fig. 3.3 Increasing the depth of the container reduces the amount of water retained and increases the air capacity of the compost.

and container depth will, of course, depend upon the pore size distribution in the compost.

3.5 FORMULATION OF COMPOSTS: PHYSICAL PRINCIPLES

From the practical viewpoint, the most important physical properties of a compost will be its total pore space, the air capacity, the amount of water available at low tensions or stress and its bulk density. It is often difficult to achieve all of the desired physical properties by using only one material and frequently mixtures are made of two or more materials. The following results of experiments at the Glasshouse Crops Research Institute show how the physical properties of a compost can be regulated by mixing two materials in varying proportions and also by regulating their grading or particle sizes.

54 Physical Aspects

RATIOS OF PEAT AND SAND

Five composts were prepared using the following volume ratios of peat to sand:

(a) 100 to 0; (b) 75 to 25; (c) 50 to 50; (d) 25 to 75; and (e) 0 to 100.

The peat was a medium grade sphagnum and the sand was a fine grade.

The bulk density, total pore space and air capacity of these composts were measured in containers 11·5 cm deep and the results are shown in Figure 3.4. Reducing the amount of peat in the compost reduced the total pore volume and increased the bulk density. In both cases the relationship was linear. The air capacity, however, showed a curvilinear reduction, as the amount of peat

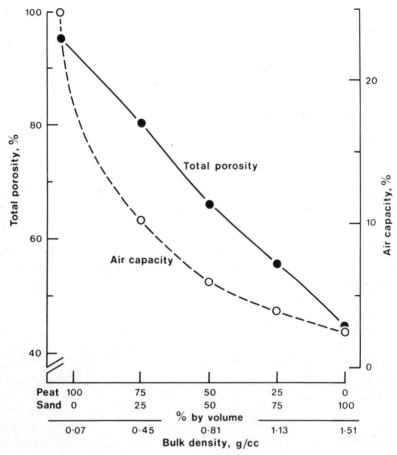

Fig. 3.4 The effect of the ratio of peat to sand on the total porosity, air capacity and bulk density of the compost.

was progressively reduced. It is possible, therefore, to prepare composts with a wide range of physical parameters by varying the peat-to-sand ratio. Partly on the basis of this data and partly from nutritional considerations, the mixture normally selected at the Glasshouse Crops Research Institute for preparing pot plant composts has 75% by volume of sphagnum peat and 25% of fine sand. This mixture has good total porosity and air capacity values, whilst the bulk density is sufficiently high to give adequate stability to tall plants, such as fuchsia and geranium, which are liable to topple over especially when light-weight plastic pots are used in place of the heavier clay pots.

GRADES OF PEAT AND SAND

Total porosity and air capacity values given in Figure 3.4 should not, however, be regarded as being absolute or intrinsic values for a particular mixture of peat and sand. The grades or particle sizes of these materials are subject to wide fluctuations and the extent to which this could affect the principal physical properties was determined by preparing composts from different grades of peat and sand. The peat used was either the normal medium grade of sphagnum peat or it was first put through a hammer mill; the sand was either a coarse grade of granite grit or the same material crushed and screened to give a very fine grade. Two series of composts were prepared from these materials, one series having 75% by volume of peat and the other series 50% by volume of peat, thereby making a total of eight composts. The grading of the peats and sands was as follows:

medium grade peat, >12 mm 24·0%, 6–12 mm 22·9%, <6 mm 53·1%
fine peat, <1 mm 100%
coarse grit, >5 mm 10·6%, 5–3 mm 42·2%, 3–2 mm 35·1%, 2–1 mm 11·5%, <1·0 mm 0·6%
fine sand, >2 mm 3·9%, 2–1 mm 9·2%, 1·0–0·5 mm 50·9%, <0·5 mm 35·6%

The principal physical parameters of these composts are given in Table 3.3 and the main effects of the grades and ratios of the materials can be summarised as follows:

Total porosity. The lowest porosity value of 64% was obtained with a mixture of 50% of fine peat and 50% of coarse grit whilst the highest porosity of 81·9% was obtained with the 75% medium peat 25% fine sand mixture. Composts made with 75% peat had approximately 11% higher porosity than composts made with 50% peat. The medium grade of peat gave about 3% higher porosity values than the finely milled peat and the fine sand gave about 4% higher porosity than the coarse grit.

Air capacity. This parameter was the one most affected by both the ratio of the ingredients and their grades. Air capacity values ranged from 2·1 to 16·6%.

Physical Aspects

Table 3.3. *Effect of the ratio of peat to sand and the grades of the materials on the physical properties of the compost.*

Parameter	Ratio (Peat:Sand)	Grade of Peat and Sand*			
		F/F	F/C	M/F	M/C
Total porosity (%)	50/50	70·8	64·0	71·4	68·4
	75/25	80·5	76·5	81·9	80·4
Air capacity (%)	50/50	2·1	10·0	5·0	15·7
	75/25	6·2	9·2	8·6	16·6
Bulk density (g cm^{-3})	50/50	0·71	0·88	0·70	0·79
	75/25	0·42	0·55	0·40	0·46
Easily available water (Air capacity to 100 cm tension)	50/50	45·7	33·5	43·4	33·7
	75/25	43·8	37·5	41·9	37·4
Reserve water (100 cm to 500 cm tension)	50/50	13·5	11·0	15·5	11·5
	75/25	17·5	16·5	18·5	15·0

*F = fine grade, M = medium grade, C = coarse grade (The particle sizes are given in the text)

Increasing the amount of peat from 50 to 75% increased the air capacity by almost 4% when a fine sand was used.. The amount of peat in the compost had no effect when a coarse grit was used. The mean air capacity value of mixtures made with fine peat was 6·9%; this increased to 11·5% for composts made with the medium grade of peat. Mixtures made with the fine sand had a mean air capacity of 5·5% and this rose to 12·9% for mixtures having the coarse grit. The particle size or grade of sand therefore had a greater effect than the grade of peat and upon the air capacity.

OVERWATERING RISK

There is less risk of overwatering seedlings and small plants in winter when the water requirements are minimal, if the compost has a high air capacity value. This was shown when two sets of tomato seedlings were grown in the eight composts already described in Table 3.2 (Bunt, 1974, a). One set of plants received the normal careful management and was given water only when necessary. The other set was given water each day, irrespective of their actual need, sufficient water being given to produce some drainage from the pot. This drainage water was collected and returned to the pot at the next watering. This minimised the development of any nutritional effects. All composts were given the same amount of fertilisers, apart from a reduced amount of lime being given to the composts made with the lower amount of peat. The results of this experiment are shown in Figure 3.5 where the weights of the plants grown in the overwatered composts have been expressed as a percentage of the weights of plants grown in the corresponding compost which

Formulation of Composts: Physical Principles 57

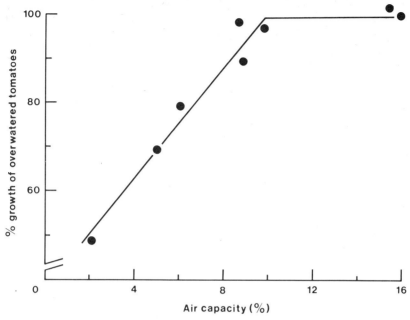

Fig. 3.5 Relationship between the air capacity of the compost and the risk of overwatering. Plants grown in composts with low air capacities and watered too frequently showed reduced growth, in comparison with plants grown in similar composts and given careful watering. When the air capacity of the compost was 10% or greater, there was no reduction in plant growth at the high frequency of watering.

had careful watering. As the air capacity was reduced below 10%, there was a progressive reduction in the amount of growth made by the plants in the overwatered composts. In the compost with only 2% air capacity the growth of the overwatered plants was only 50% of plants in the same compost receiving correct watering. At air capacities above 10% there was no reduction in growth by overwatering. This experiment shows the importance of using a compost with a high air capacity whenever there is the risk of plants being overwatered.

Certain plants are considered to be more susceptible to overwatering and lack of aeration than others and examples of plants requiring differing air capacities are given by Johnson (1968) (Table 3.4).

Easily available water. The general effect of the ratio of the ingredients and of their particle size or grade on the amount of water available between container capacity and 100 cm water tension was the reverse of that found with the air capacity. When composts were made with only 50% of peat, using a coarse grit instead of a fine sand, the easily available water content was reduced by about 11%. When the peat content was increased to 75% by volume, the effect of the grade of sand was reduced; the coarse grit gave about 5% less easily available water than the fine sand. The grade of peat, however, had very little effect upon the amount of easily available water.

Table 3.4. *Approximate root aeration requirements of selected ornamentals, expressed as the free porosity (Johnson, 1968)*

Very high 20	High 20–10	Intermediate 10–5	Low 5–2
Azalea	Antirrhinum	Camellia	Carnation
Orchid	Begonia	Chrysanthemum	Conifer
(Epiphytic)	Daphne	Gladiolus	Geranium
	Erica	Hydrangea	Ivy
	Foliage plants	Lily	Palm
	Gardinia	Poinsettia	Rose
	Gloxinia		Stock
	Orchid		Strelitzia
	(Terrestrial)		
	Podocarpus		Turf
	Rhododendron		
	Saintpaulia		

COMPACTION

In addition to the effect which the grades and ratios of the compost ingredients have upon the physical parameters of the compost, the degree of compaction or the amount of compression and firming given during potting is another important factor. It has been traditional practice to pot certain plants firmly. For example, when the winter-flowering chrysanthemum was given its final potting, it was customary to use a 'rammer' or wedge-shaped potting tool to achieve firmness or compaction of the compost. The principal effects of firm potting are a reduction in the total porosity and therefore a reduction in the amount of available water. Both the quantity of compost required to fill the pot and the amount of soluble nutrients in the pot are increased. This reduction in the amount of water, coupled with the increase in the amount of soluble salts, means that there is a significant increase in the solute or osmotic stress. The net result of this is to increase the total soil moisture stress and this produces a slower growing and harder type of plant.

Modern cropping schedules call for a quick-growing plant and it is customary to apply little or no firming during potting. The physical data given in this chapter are based on composts at their 'natural' densities, i.e. loosely filled containers in which the compost has been compacted to its natural density by applying water with a 'rose' or spray.

Much of the discussion on the physical aspects of composts and their formulation has been with reference to mixtures of peat and sand. By using various combinations of the wide range of materials available, it is possible to formulate an infinite number of composts, and it would not be practical to attempt a discussion of their individual physical characteristics. The physical principles relating to the formulation of peat-sand compost will, however, apply equally to composts made with other materials. White and Mastalerz (1966) made a detailed investigation of the principal physical properties of thirty composts based on soil-perlite-peat mixtures. Waters *et al.* (1970) have examined twenty-seven composts made from mixtures of builders' sand, wood shavings, perlite and two peats, and Joiner and Conover (1965) examined the physical properties of twenty composts made from sand, perlite, vermiculite, peat and bark. All these studies have shown that composts with suitable physical properties can be prepared from combinations of many different materials.

Chapter 4

Principles of nutrition

The principal requirements for plant growth are adequate levels of light, temperature, and carbon dioxide in the aerial environment, and of water, air and mineral nutrients within the compost. It is the various aspects of mineral nutrition with which we are now concerned.

The leaf and stem tissue of a typical pot plant can be regarded as consisting of 90% water and 10% solids; these values will vary a little with the age of the plant, the species and the general environment. The dry matter can next be subdivided into organic and inorganic fractions consisting of approximately 90% organic compounds such as cellulose, sugars, proteins and enzymes, etc., with the remaining 10% being minerals such as calcium, potassium, magnesium, phosphorus and micro elements. The amount of dry matter produced by a pot plant and hence the total amount of minerals required, varies considerably, ranging from about 30 g for a pot chrysanthemum grown in a 14 cm pot holding one litre of compost, to 3 g for a Saintpaulia grown in a 8 cm pot holding 300 cc of compost; the dry weight of a box of bedding plants grown in 4·5 l of compost would be about 20 g. Mineral composition of the dry matter also varies slightly with the plant species and the nutrition it receives, but the following would be typical values of minerals found in the leaves of the average pot plant:

Nitrogen 3–4·5%, Phosphorus 0·3–0·6%, Potassium 3–4·5%
Calcium 1–2%, Magnesium 0·2–0·5%.

Micro elements such as copper and boron are present in much smaller concentrations, e.g. 10 to 100 ppm. Although the minerals represent only about 1% of the total fresh weight of the plant, they are nevertheless essential if the various growth processes are to function efficiently. Adequate levels of minerals must therefore be maintained in the compost at all times if maximum rates of growth are to be achieved.

4.1 CATION EXCHANGE CAPACITY

Plant nutrients are normally applied to the compost as salts, which are

Cation Exchange Capacity   61

composed of atoms carrying electrical charges known as ions; those with a positive charge are called *cations* whilst those with a negative charge are called *anions*. The salt potassium nitrate (KNO_3) comprises potassium (K^+) cations and nitrate (NO_3^-) anions.

One of the important mechanisms which help to regulate the supply of certain nutrients to the plant is known as the cation or base exchange capacity (CEC). This is defined as the sum of the exchangeable cations or bases that a soil can absorb per unit weight and is usually expressed as milligram equivalents per 100 g; this is sometimes abbreviated to me or meq/100 g. Cation exchange capacity is normally associated with the clay particles in a mineral soil, and whilst organic matter such as humus and the

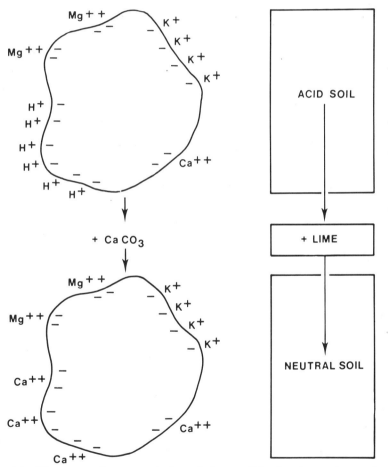

Fig. 4.1   Diagrammatic representation of clay particle showing cation exchange process when calcium carbonate is applied to the soil.

roots of plants also have cation exchange capacities, it is convenient for the present to follow the convention of considering this phenomenon in relation to mineral soils. Each clay particle has a number of negative charges on its surface and when fertilisers are applied to the soil the cations such as calcium^{++}, magnesium^{++} and hydrogen$^+$ are attracted to the clay particle. The cation exchange process is illustrated in Figure 4.1 by showing the reaction which occurs when lime in the form of calcium carbonate or ground lime is applied to an acid soil; this process also occurs with other salts or fertilisers. The cations commonly associated with plant nutrition are Ca^{++}, Mg^{++}, K^+, NH_4^+ and Na^+, and these have been arranged in order of decreasing retention by the clay particle, i.e. sodium and ammonium are not as strongly retained as magnesium and calcium. The total cation exchange capacity is not confined solely to the clay; other minerals such as vermiculite, and organic matter such as peat also exhibit this phenomenon. The cation exchange capacity of some of the materials used in compost making is given in Table 4.1.

The CEC of a material normally increases with the pH at which the

Table 4.1. *Cation exchange capacities of some materials and composts.*

	Cation exchange capacity	
Materials and composts	*me/100 g*	*me/100 cc*
(1) Humus	200	
Vermiculite	150	
Montmorillonite	100	
Illite	30	
Kaolinite	10	
(2) Fine clay	56–63	
Coarse clay	22–52	
Silt	3–7	
(3) Perlite	1·5	
(4) Peat, sphagnum	100–120	
Composts		
(5) 50% vol. sphagnum peat 50% sand	8·01	5·35
75% vol. sphagnum peat 25% sand	18·16	6·75
50% vol. sedge peat 50% sand	21·11	17·12
75% vol. sedge peat 25% sand	40·05	21·80
John Innes compost	8·80	7·70
(6) 25% peat 75% Perlite	11·2	1·0
50% vermiculite 50% peat	140·96	32
66% pine bark 33% perlite	23·58	5

(1) Thompson, L. M. (1957); (2) Whitt, D. M. and L. D. Baver (1930); (3) Merrison, T. M., D. C. McDonald and Jean A. Sutton (1960); (4) Puustjärvi (1968); (5) Bunt, A. C. and P. Adams (1966); (6) Joiner, J. N. and C. A. Conover (1965).

determination is made and the standard procedure is to report the value for pH 7. In the case of peat-based composts which are normally used in the range pH 5 to 6, the CEC value will in practice be significantly lower than the values given in Table 4.1. Whilst pot plants can be grown satisfactorily in a wide range of materials irrespective of their cation exchange capacities, management and plant nutrition are always easier when the compost has a reasonable exchange capacity.

4.2 ANION EXCHANGE CAPACITY

In addition to the cations, plants also require the negatively charged anions, such as nitrate (NO_3^-), chloride (Cl^-), sulphate (SO_4^{--}) and phosphate ($H_2PO_4^-$). Whilst mineral soils have a permanent negative charge and so attract cations in the manner described, they do not possess a similar attraction for anions. Some phosphate anions are, however, adsorbed by the clay particles and can act as replaceable anions. The pH or acidity of the soil has a large effect on the forms and availability of phosphorus to the plant. When a water soluble form of phosphorus such as superphosphate is added to a mineral soil, it is quickly converted from the soluble monocalcium phosphate to the insoluble dicalcium phosphate. Under strongly acid conditions, the phosphorus combines with the aluminium and iron to give strongly fixed or insoluble aluminium and ferric phosphates. These insoluble phosphates become slowly available to the plants by the action of the weak acids present in the soil solution. Mineral soils usually contain only very low concentrations of phosphorus in a water soluble form and this would be quickly depleted by the plant, unless it was continuously renewed from the insoluble or reserve forms of phosphorus present in the soil. It has been estimated that the amount of phosphorus released in this manner is equivalent to the complete replacement of the water-soluble phosphorus several times each day.

Unlike the mineral soils, peats do not have a significant anion exchange capacity and they contain relatively little iron and aluminium. Probably the aluminium is of greater importance than iron in the fixation of the relatively small amount of phosphorus which does occur in peats. Consequently a large part of the phosphorus applied as superphosphate to the peat remains in the water soluble form and is readily available to plants; one disadvantage of this, however, is that it can be easily leached from the composts.

4.3 AVAILABILITY OF NUTRIENTS: LOAM V. LOAMLESS COMPOSTS

CHEMICAL ANALYSIS

The difference in the behaviour of the principal plant nutrients when applied to a peat–sand compost and a mineral soil compost (John Innes) is shown in Table 4.2. The nitrogen, phosphorus and potassium levels in these two composts were determined both before and after adding a base fertiliser which supplied known amounts of ammonium, nitrate, phosphorus and potassium; the peat–sand compost also had the normal amount of calcium carbonate and Dolomite limestone added. The availability of these nutrients

Table 4.2. *Percentage of added nutrients found by chemical analysis of a mineral soil compost (John Innes) and peat–sand compost (based on water extracts and Morgan's solvent).*

	John Innes		Peat–sand	
Element	Water	Morgan's	Water	Morgan's
NH_4–N	69	88	79	90
NO_3–N	81	90	87	95
P	9	16	56	66
K	40	57	74	91

was then assessed by keeping the compost moist but avoiding leaching for two weeks to allow any fixation of the nutrients to occur. The composts were then extracted with two extractants, viz. distilled water and Morgan's solvent, the latter being 0·52N acetic acid buffered with 0·73N sodium acetate and having a pH of 4·8. This and similar type of weak acid extractants are generally regarded as giving some measure of the availability of nutrients to plants.

With the John Innes compost, 16% of the added phosphorus was extracted by the weak acid and 9% by the water, whereas in the peat-sand compost the corresponding values were 66% and 56%. The weak acid had extracted 80% more phosphorus than the water had from the John Innes compost but only 20% more than the water had from the peat–sand compost. The much greater water-solubility of phosphorus in the peat–sand compost means that it is less likely to become a limiting factor in plant nutrition, *providing that it is not lost by leaching.* Potassium was also more easily removed by both extractants from the peat–sand compost than it was from the John Innes compost.

Other anions, such as chloride and nitrate, are retained by neither the mineral nor the organic fractions of composts, and are readily leached from all materials. The plant must therefore rely upon nitrate nitrogen lost by leaching being replaced with nitrogen made available from either organic sources and other forms of slow release fertilisers, or from the regular application of liquid feeds. Nitrogen can also be present in composts as a cation in the ammonium form. This cation is not, however, as strongly retained as calcium and magnesium and in contrast to field conditions, a considerable amount of ammonium can be lost from composts by leaching.

Some experiments have been made on the use of mixed ion exchange resins in composts. These help to maintain a supply of both cations and anions to the plant but as indicated earlier there is not yet sufficient experience on which to base positive recommendations for their use.

PRESENTATION OF ANALYTICAL RESULTS

Usually a weight measurement of compost can be made more precisely than a volume measurement, and for this reason the weight basis has been traditionally used in soil analysis. Whereas mineral soil composts, e.g. John Innes, normally have a bulk density in the pot of about 1 gram cm^{-3}, composts based on light-weight materials such as vermiculite, perlite, peat and peat-sand mixtures often have bulk densities in the range 0·1 to 0·5 g cm^{-3}. Comparison of the results of an analysis of these different composts made on a weight basis would therefore be very misleading, the light-weight compost would appear to contain much more nutrients than the heavier composts. To avoid this error, either a bulk density determination of each compost must be made and the analytical results corrected before presentation, or alternatively the chemical analysis must be made on a standard volume of compost. The volume basis is now generally accepted as more meaningful and relevant in the case of potting composts than the traditional weight basis of presentation. Plants are, of course, grown in a *volume* rather than a *weight* of compost, and a comparison of the cation exchange capacities for those materials and composts given in Table 4.1, on both the weight and the volume basis, shows how misleading an uncorrected weight-based presentation could be. The same situation applies to a mineral analysis for the individual cations and anions, i.e. NH_4, NO_3, P, K, Ca and Mg, and all analytical results should be expressed as milligrams per litre of compost.

LOSS BY LEACHING

The relatively high frequency of irrigation required by container-grown plants has already been discussed. Inevitably, this must result in some loss of water from the containers, together with the possible loss of nutrients. In view of the high solubility of most minerals in the peat–sand composts as already shown in Table 4.2, the rate at which minerals could be lost from loamless composts made from different materials was examined under typical growing conditions (Bunt, 1974, *a*). Known amounts of nutrients were first added to individual pots each holding one litre of compost. Sufficient water was then given at a high rate of application, i.e. equivalent to hose watering, to collect leachate in increments of 50, 75, 125, 250 and 500 ml, thereby making a total of one litre of leachate from each pot. This was equivalent to the total volume of each pot, or an irrigation equivalent to

66 *Principles of Nutrition*

Table 4.3. *Percentage of added nutrients recovered in one litre of leachate from peat-sand (75%–25%) and peat-vermiculite (50%–50%) composts. Pot capacity – 1 litre.*

Element	Peat-sand	Peat-vermiculite
NH_4-N	81	33
NO_3-N	87	75
P	60	43
K	70	45

6·5 cm depth of water. Two sets of composts were prepared, one consisting of 75% by volume of sphagnum peat and 25% of fine sand, the other of equal parts of sphagnum peat and vermiculite.

The total amounts of nutrients recovered from these composts are given in Table 4.3 and the rate at which they were leached from the composts is shown in Figures 4.2 and 4.3. All nutrients showed an exponential rate of loss, with 28% or more of the nutrients added to the peat–sand compost

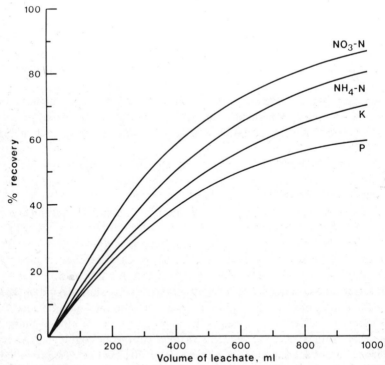

Fig. 4.2 Showing the rate at which nutrients are leached from a peat-sand compost. (Compare with Fig. 4.3.)

being lost in the first 250 ml of leachate. Apart from nitrate nitrogen, the peat–vermiculite compost showed a much lower rate of nutrient loss, only 15 to 22% of the nutrients being lost in the first 250 ml of leachate. In addition to its known ability to hold potassium and ammonium in an exchangeable form, the vermiculite also reduced the amount of phosphorus lost by leaching.

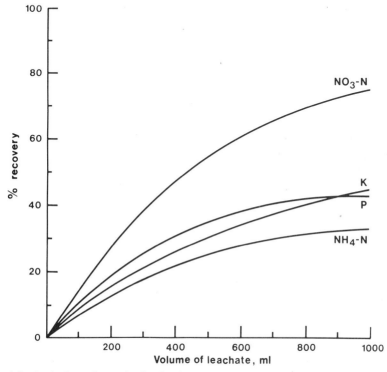

Fig. 4.3 Inclusion of vermiculite in the compost significantly reduced the rate of nutrient loss, apart from the nitrate nitrogen.

4.4 NUTRIENT UPTAKE BY THE PLANT

Plants absorb nutrients in the form of cations and anions through their root hairs which are covered by a film of water containing dissolved nutrients. This is sometimes referred to as the soil solution. The concentration of nutrients found in plant roots, however, is many times greater than that present in the soil solution and it is evident that diffusion cannot play an important part in nutrient absorption. Nutrient absorption is an energy consuming process; sugars that have been manufactured in the leaves are transported to the roots where they are used to provide the energy required for root respiration and nutrient uptake. Ions are absorbed selectively and

68   Principles of Nutrition

independently by the plant. For example, if a fertiliser such as ammonium sulphate is added to the compost, the plant will absorb many more NH_4^+ ions than SO_4^{--} ions; similarly when calcium nitrate is added, more NO_3^- ions will be absorbed than Ca^{++} ions. The rate of ion absorption is dependent upon several factors other than their actual concentration in the soil solution. A low supply of sugars in the plant, low soil temperature and low oxygen supply will all reduce the rate of absorption. Plants also release ions through their roots; for example, hydrogen can be released in exchange for calcium, and bicarbonate ions can be released in exchange for nitrate ions.

Frequently, the rate of plant growth can be increased by increasing the supply of nutrients in the compost, but it is essential that a balance between the nutrients is maintained at all times. If, for example, the amount of N, P and K available to the plant is increased, it is quite possible that magnesium deficiency symptoms will appear. This will have been caused partly by the higher growth rate and the greater demand for magnesium, and partly by the antagonism which exists between potassium and magnesium; high potassium levels actually render magnesium less available. This is one of the best known instances of antagonism in plant nutrition but several other examples exist. The reverse case of potassium deficiency in chrysanthemum, induced by the use of fertilisers very high in magnesium, has been observed in experiments with loamless composts made at the Glasshouse Crops Research Institute. Some ions, however, increase the absorption of others, e.g. the anion NO_3^- is more effective than other anions in increasing the uptake of the cations Ca^{++} and K^+; the NO_3^- anion also reduces the uptake of the

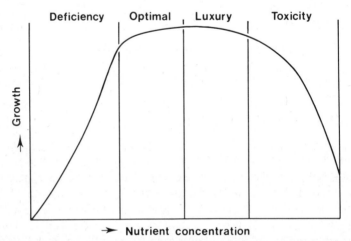

Fig. 4.4 The relationship of the rate of plant growth with the nutrient concentration in the compost. Once the optimal nutrient level is reached, there is the risk of suppressing growth.

SO_4^{--} anion. Some minerals will also partially substitute for others in the plant: sodium, for example, will reduce the need for potassium.

Whilst an increase in the general supply of nutrients will often give an increased rate of growth, it is possible to arrive at a situation whereby growth has been depressed by too great a supply of nutrients. This change in the response of the plants to the supply of nutrients is represented diagrammatically in Figure 4.4. Usually the plant does not absorb nutrients at a uniform

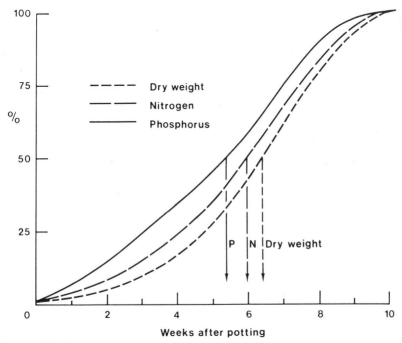

Fig. 4.5 Typical curves showing the rate of dry weight accumulation and nutrient uptake for pot chrysanthemums. Phosphorus was taken up at a high rate by the young plants.

rate over its growing period: when plants are young and growing actively they often contain higher concentrations of nutrients than older, slower growing plants. A typical pattern of dry matter accumulation and nutrient uptake is shown in Figure 4.5. At four weeks after potting, the chrysanthemums had made 16% of their final dry weight but had already absorbed 35% of the final phosphorus uptake. Also nutrients are not always distributed evenly within the plant. Plants grown in composts with high levels of copper have been found to have much higher concentrations of copper in their roots than in their shoots.

4.5 ACIDITY (pH)

Materials used in making composts differ widely in their pH or acidity; peats are usually acid in reaction whilst some of the vermiculites are alkaline. The degree of acidity or alkalinity is customarily expressed as a pH value, which is a measure of the relative concentrations of hydrogen (H^+) and hydroxyl (OH^-) ions in solution. The pH scale ranges from 0 to 14, with pH 7 being the neutral point where the number of hydrogen ions balance the hydroxyl ions. When there are more hydrogen ions than hydroxyl ions the material is acid and the pH will be below 7. Conversely, when the hydroxyl ions exceed the hydrogen ions the material will be alkaline with a pH value above 7. Technically speaking, the pH is the common logarithm of the reciprocal of the hydrogen ion concentration, and as the pH values are in a logarithmic scale this means that pH 5 is ten times as acid as pH 6, whilst pH 4 is 100 times as acid as pH 6.

pH MEASUREMENT

This can be made either with indicator solutions, whose change in colour is related to the hydrogen ion concentration, or potenticmetrically with a pH meter and a glass electrode system. There are a number of indicators that can be used to measure the pH of a compost. Bromo-cresol green changes from yellow at pH 3·6 to blue at pH 5·2, chlorophenol red changes from yellow at pH 4·6 to violet at pH 7·0, and bromo-thymol blue changes from yellow at pH 6·0 to blue at pH 7·6. To overcome the relatively narrow bands over which these indicators work, a universal indicator with a range from pH 3·0 to pH 11·0 is available, but there will be some loss of accuracy as readings cannot be estimated closer than 0·5 of a pH unit. A wide-range indicator can be prepared from a mixture of two parts of bromothymol blue and one part of methyl red. The range of colour changes in relation to the soil pH are:

Colour	pH
Brilliant red	<3·0
Red	3·1–4·0
Red-orange	4·1–4·7
Orange	4·8–5·2
Orange-yellow	5·3–5·7
Yellow	5·8–6·1
Greenish-yellow	6·2–6·4
Yellowish-green	6·5–6·7
Green	6·8–7·3
Greenish-blue	7·4–7·8
Blue	>7·9

When this indicator is used on some peats, however, a preferential adsorption of one of the dyes can cause misleading results. Indicator solutions are useful for quick tests, but they cannot give the accuracy of a properly conducted test using a pH meter.

Erroneous readings can, however, be obtained with a pH meter if the test is not made correctly. The conventional method of making a pH test on a suspension of 1 part of soil or compost to $2\frac{1}{2}$ parts of distilled water may give pH readings that are significantly higher than those experienced by plants actually growing in the compost. This increase in the pH reading can be as much as 0·5 to 1·0 pH unit. When a suspension is prepared using a wide ratio of compost to water, the salts present in the compost are diluted and the pH increases: this is known as the soil–water ratio effect. To obtain a pH which is representative of that experienced by the plant in the pot, the amount of water used should not be greater than that found under normal growing conditions. Measurements at this water content can only be made by using a specially designed 'spear' electrode; the normal thin-walled glass electrode would fracture if used under such conditions. An alternative procedure is to make a suspension using a weak electrolyte solution such as 0·01 Molar calcium chloride in place of the distilled water. This overcomes errors caused by the soil–water ratio effect and avoids the errors when determining the pH of composts containing high levels of ammonium; this is a situation which frequently occurs when slow release forms of nitrogen are used or when urea is included in the liquid feed. For example, pH determinations made with a spear electrode on two composts, one of which had no ammonium present and the other 90 mg/l of ammonium nitrogen, showed that in a moderately moist condition, similar to that occurring in a pot sometime after it had been watered, the pH of the two composts was similar at 5·5. When conventional suspensions of 1 part compost to $2\frac{1}{2}$ parts distilled water were used for the pH determinations, the pH of the compost without any ammonium rose to 5·8, an increase of 0·3 units, whilst for the compost containing ammonium it rose to 6·3, an increase of 0·8 units. When M/100 $CaCl_2$ was used in place of the distilled water, the pH of both composts was approximately 5.5 (Bunt and Adams, 1966a).

MINOR ELEMENT AVAILABILITY

The pH value of the compost is important for a number of reasons. Whilst at very low pH values the hydrogen ions can themselves cause toxicity, it has been shown that plants are able to grow over a wide range of hydrogen ion concentration, from pH 4 to pH 8, before the acidity or alkalinity as such causes any trouble, *providing that the minor elements required by the plant are maintained in an available form* (Arnon and Johnson, 1942). Usually the pH will be of concern to the grower and the soil chemist because of its effects on the availability of plant nutrients, not all of which behave in the same manner. Whereas interactions of nutrient availability and the pH in mineral soils and composts have been studied for many years, it is only relatively recently that much experience has been gained with respect to organic materials. Lucas and Davis (1961) in the USA have reviewed work on nutrient availability in organic soils and have summarised this with a chart, reproduced in Figure 4.6, showing how the availability of the twelve most important elements changes in relation to the pH. They concluded that the optimal pH value for organic soils and peats was in the range pH 5·0 to 5·5;

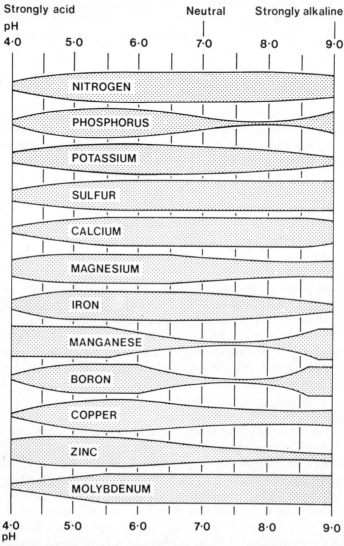

Fig. 4.6 The effect of the soil pH on the availability of plant nutrients in organic soils. (From Lucas and Davis, 1961.) The availability of the elements is indicated by the width of the bands.

this is about 1·0 to 1·5 pH units lower than the corresponding values for mineral soils, including the John Innes composts.

4.6 LIME REQUIREMENT

Whilst a pH determination measures the *active* soil acidity it does not give a direct indication of the *reserve* or *exchange* acidity. It is possible for two materials to have the same pH or active acidity, but significantly different amounts of reserve acidity. They will, therefore, have different lime requirements. Composts containing large quantities of sand, perlite and expanded plastics, for example, will require only relatively small quantities of lime, whilst composts made from peats and mineral soils having a high clay content will require much more lime to effect a given increase in the pH. Such materials are said to have a high lime requirement and to be well buffered against a change in pH.

The lime requirement of four composts made from 75% by volume of peat and 25% of sand is shown in Figure 4.7; three of the peats were of the sphag-

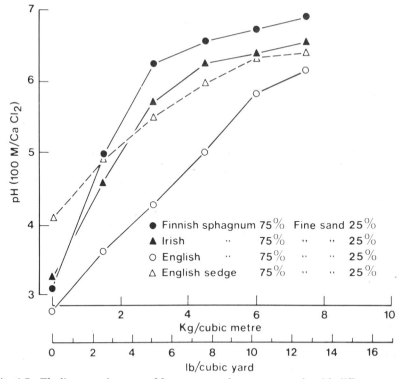

Fig. 4.7 The lime requirement of four peat-sand composts made with different peats. The liming material was equal parts of calcium carbonate and Dolomite limestone.

74   *Principles of Nutrition*

num type, originating from different countries, and the fourth was a sedge peat. The pH of the sedge peat compost, before any lime had been added, was about 1 pH unit higher than any of the sphagnum peat composts, but its rise in pH with increasing amounts of lime was relatively low. Whilst the initial pH of the three sphagnum peat composts, before any lime had been added, was very similar for all peat sources, the Finnish peat compost had a slightly lower lime requirement than the Irish peat compost, and the English sphagnum peat compost had a very high lime requirement, almost twice as much lime being required to reach pH 6 as for the other two sphagnum composts. For practical purposes, the lime requirements over the linear part of the curves for composts made from these different peats are:

Compost	Lime required to raise the pH 1 unit
75% Finnish sphagnum 25% Fine sand	0·9 kg/cu.m (1 lb. 8 oz./cu.yd)
75% Irish sphagnum 25% Fine sand	1·1 kg/cu.m (1 lb. 14 oz./cu.yd)
75% English sphagnum 25% Fine sand	2 kg/cu.m (3 lb. 6 oz./cu.yd)
75% English sedge 25% Fine sand	2 kg/cu.m (3 lb. 6 oz./cu.yd)

These results show the need to know the exact lime requirement of each compost rather than work to rule-of-thumb measurements with unknown materials.

Using Irish sphagnum peat, the effect of the ratio of peat to sand in the compost is shown in Figure 4.8, and the lime requirement over the linear parts of the curves can be said to be proportional to the peat content of the compost.

In both of these experiments the 'lime' consisted of equal quantities of calcium carbonate and Dolomite limestone. To avoid any extraneous effects, other fertilisers were not included in the composts. The normal base fertilisers used in the GCRI peat–sand compost (p. 156) would have reduced the pH by approximately 0·5 units. This is mainly due to the acidifying action of the superphosphate (Table 6.1); other fertiliser mixtures could be expected to give slightly different results. All pH determinations were made after an equilibrium between the lime and the moist compost had been reached, and the pH values refer to determinations made on a compost suspension in $M/100$ $CaCl_2$.

4.7   SOLUBLE SALTS

To achieve the high rates of growth required in modern pot plant culture, the levels of the major nutrients in the composts are maintained at relatively high concentrations by using large amounts of base fertilisers and high strengths of liquid fertilisers. To some extent this practice compensates for the small volume of compost available to the plant roots and the rapid depletion of nutrients that would otherwise follow. We have already seen that the nutrients required by plants are present in varying degrees in the

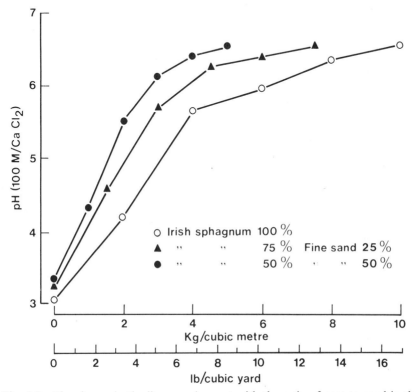

Fig. 4.8 The change in the lime requirement with the ratio of peat to sand in the compost.

soil solution, and whilst an increase in their concentration will often result in an increased growth rate, very high concentrations can restrict growth or even cause the death of the plant. This build-up in nutrients, which is often referred to as a high salinity, operates either as a specific ion toxicity or as a general salinity effect by reducing the availability of the water to the plants. Specific ion toxicities, such as those of manganese or boron, occur only infrequently, whereas a high level of soluble nutrients, principally of nitrate and potassium in potting composts, can occur whenever fertilisers are applied in excess of the rate of plant uptake and the loss by leaching.

Principles of Nutrition

PLANT RESPONSE

Water stress within the plant is one of the factors which directly control plant growth. The magnitude of the stress is controlled by:

(1) the water stress within the aerial environment, i.e. the vapour pressure deficit, which affects the transpiration rate;
(2) the water stress within the compost which affects the rate of water absorption by the plant.

It is the latter with which we are now concerned.

We have already seen that the total soil moisture stress comprises the matrix tension and the solute stress. Both of these components affect plant growth by reducing the availability of the soil water and may thereby produce a water deficit within the plant. A number of workers have demonstrated that salts dissolved in the soil solution can reduce the availability of the water to plants. Eaton (1941) used the 'split-root' technique to show that when the roots of corn plants were equally divided between two nutrient solutions, one with a salt stress of approximately $\frac{1}{3}$ of an atmosphere and the other $1\frac{3}{4}$ atmospheres, the plants took up water from the weak solution at approximate twice the rate at which water was absorbed from the stronger solution. It can also be shown that, under certain conditions, the effects of matric tension and solute stress in soils are additive in their effects on plant growth. Wadleigh and Ayers (1945) grew bean plants in soil cultures which were allowed to dry out to different soil water tensions before they were watered. Within each set of water tension treatments, a range of osmotic stress treatments was obtained by adding different amounts of sodium chloride to the soils. Their results, reproduced in Figure 4.9, show that the plants integrated the two separate factors, and responded to the total soil moisture stress irrespective of its composition. Subsequent work has shown, however, that this concept may be an oversimplification. Plants are able to adjust to an increase in the soil solute stress by making a corresponding increase in the osmotic pressure of their cells. Whilst a sudden increase in the salinity of the soil can cause the plants to wilt, they are usually able to adjust to the higher soil salinity at a rate of about one atmosphere per day (Bernstein, 1963). Turgidity will then be regained, unless the roots have received permanent damage. Another possible cause of the reduction in plant growth, observed under saline conditions, is the specific ion toxicity effect. For example, Gauch and Wadleigh (1944) found, when growing beans in water cultures of varying solute stress induced by different salts, that growth was inhibited more by magnesium sulphate and magnesium chloride than it was by sodium sulphate and sodium chloride solutions of a similar solute stress. Thus the growth of plants in soils with a high solute stress will be reduced because soil water is less available, and growth may be further reduced by a

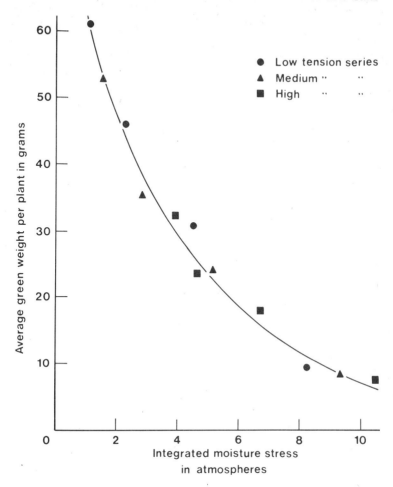

Fig. 4.9 Plant growth responds to both water and salt stress and these components have been integrated into the total soil moisture stress. (From Wadleigh and Ayers, 1945.)

specific ion toxicity. The response of the plant to soil salinity will also be determined by the aerial environment in which it is growing. When the vapour deficit is high, as occurs in a glasshouse in summer under conditions of high solar radiation, high temperature and low humidity, the reduction in growth due to a high soil moisture stress will be much greater than in winter, when the vapour pressure deficit is less and the rate of transpirational loss is lower.

What are the measures which the pot plant grower can adopt to control the salinity of the compost and to reduce its detrimental effects on plant growth?

78   Principles of Nutrition

(1) The compost should be kept moist. This will not only reduce the matric tension, it will also prevent the consequent increase in the solute stress associated with the reduced water content (p. 49).
(2) Never apply either a powdered fertiliser or a strong fertiliser solution to a dry compost. Liquid fertilisers should be applied 'little and often'.
(3) Check the accuracy of the dilutor regularly, by measuring the EC of the liquid feed (p. 193).
(4) Avoid fertilisers that give a high solute stress for a given amount of plant nutrients (Table 6.5).
(5) When the compost salinity is high, leach with plain water.
(6) Keep plants shaded and also raise the humidity by damping down.

Symptoms of high salinity stress in plants vary with the crop, the most common signs being a reduction in the growth rate or stunting of the plant. The general appearance of the plants can range from a dark bluish colour to yellowing of the leaves or chlorosis, with occasionally necrosis or leaf burns. If the roots have been damaged the leaves may wilt even when the compost is wet.

SALINITY MEASUREMENT

It is important, therefore, that routine checks be made of the amount of salts present in the compost. There are several ways of doing this. A direct determination of the osmotic pressure of the soil solution by the method of freezing point depression gives the most accurate and fundamental measurement. It is, however, a very laborious procedure and is used only in certain experimental work. Alternative methods based upon the measurement of the electrical conductivity (EC) are usually used because of their greater speed, the EC being directly dependent upon the concentration of the dissolved salts.

To give a meaningful estimate of the salinity of the compost or the soil solution in which the plant is growing, the amount of water used for the determination must be standardised and be related preferably to the amount of water which is present in the compost under certain specified conditions. Several experimental workers have confirmed that salinity determinations based upon the saturated paste extract, i.e. EC_e (Richards, 1954), give the most accurate and meaningful results in terms of plant response.

Saturated paste extract (EC_e). This is prepared by adding distilled water to a sample of the compost and stirring vigorously; when the compost will not hold any more water, it is allowed to stand overnight and final adjustments are made the following morning. When the saturation point is reached, the compost will slide freely from a spatula, it will glisten and reflect light but there will be no free water on the surface. To make the salinity determination, the water and dissolved salts are withdrawn from the compost with a vacuum or water pump, and the extract is placed in tubes in a water bath at 25°C. The electrical conductivity is measured with a conductivity meter or Wheatstone bridge, and the results are expressed in millimhos/cm, the mho being the unit of conductivity and the reciprocal of the ohm which is the unit of resistance.

The saturated paste method of salinity determination has the great advantage that the amount of water present in the paste is related to the amount of water in the compost at container capacity, i.e. the difference in the texture and water retention of the various materials is largely overcome. For potting composts high in organic matter, the ratio:

$$\frac{\% \text{ of water present in the saturated paste}}{\% \text{ water at container capacity}}$$

or as it is commonly called, the moisture factor, ranges from 1·0 to 1·2. This means that the salinity readings have a direct relationship to the salinity experienced by the plant, and for practical purposes, the salt stress or osmotic concentration in the compost can be derived from the following formula:

osmotic concentration = $0.365 \times EC_e$ (millimhos/cm at 25°C)
(in atmospheres)

The factor of 0·365 was derived from numerous determinations of the EC and the osmotic pressure of various salt solutions by the USDA Salinity Laboratory (Richards, 1954). The moisture factor for agricultural soils is usually accepted as being 2·0. The reason for the lower values found in potting composts is the increased amount of organic matter present and the greater amount of water retained in shallow containers at 'container capacity', by comparison with agricultural soils at 'field capacity'.

Plant tolerance to salinity varies appreciably between species, and although, at present, there is more information available on the response of field and vegetable crops than of ornamentals, plants such as carnation and portulaca can be regarded as being tolerant to high salinity values. Plants such as chrysanthemum, fuchsia, and cyclamen have a medium degree of tolerance, whilst camellia and gardenia have a low salt tolerance; azaleas are especially susceptible to saline conditions. In addition to this classification it must be remembered that young seedlings are generally less tolerant than more mature and established plants. A suggested range of salinity values in relation to the degree of plant tolerance is given in Table 4.4.

EC determinations based on suspensions. This method is often used for advisory purposes rather than saturated paste extracts, because of the ease of preparing the suspension and the greater speed when dealing with large numbers. Usually a ratio of one part compost to two parts water is used, and

Table 4.4. *Interpretation of conductivity readings based upon saturated paste extracts, EC_e, millimhos/cm, 25°C.*

Plant tolerance	Desirable range (millimhos/cm)
Low	1–2
Medium	2–4
High	4–6

occasionally a wider ratio of five parts of water is used. Unfortunately this method lacks the accuracy of the saturated paste method, due in part to the big difference found in the bulk densities and water retention of the different composts. Also the quantity of calcium sulphate in solution is often affected by the compost-to-water ratio. The effect of varying compost densities can be minimised by using a *volume* rather than a *weight* ratio of compost to water. Waters *et al.* (1972) further recommend that the compost sample should be air-dry rather than moist.

Often old glasshouse border soils, that have been given fertilisers over a long period of time, contain more calcium sulphate than can be dissolved in the soil solution. When a suspension with a wide compost to water ratio is used for a salinity determination, some of the undissolved calcium sulphate then passes into solution. Thus under the conditions of the test, more dissolved salts will be present than occur under natural conditions. Calcium sulphate has a solubility of approximately 0·24 g/100 ml and a specific conductivity of 2 millimhos/cm.

In England, chemists in the Agricultural Development and Advisory Service (ADAS) use a suspension method of salinity determination based upon the work of Winsor *et al.* (1963). The variable effect of the calcium sulphate in the soil is removed by preparing a soil-water suspension to which an excess of calcium sulphate has been added (0·5 g/100 ml). The results are further corrected for the texture of the soil, using a bulk density determination.

Salinity values obtained by the suspension methods therefore give relative rather than absolute values. The range of values obtained by the suspension methods, and their interpretation are given in Tables 4.5 and 4.6. As readings obtained by these methods cannot be directly related to the results of saturated paste (EC_e) determinations, separate interpretation scales must be used (Rutland, 1972). High salinity levels usually have the greatest adverse effect on plant growth under conditions of maximum water requirement, i.e. salinity has a greater effect on plant growth in summer than in winter.

It must be remembered that with the electrical conductivity methods only the amount of electrolytes in solution is measured. The non-electrolytes, which also contribute to the osmotic pressure but which are not detected by this

Table 4.5. *Interpretation of the conductivity of a 1:2 air-dry compost to water suspension, volume basis, millimhos, 25°C.*

Peat and low bulk density composts			
Salt tolerance of plants			
Sensitive	Medium	Salt rating	Remarks
<0·37	<0·5	Low	Fertiliser needed
0·37–0·75	0·5–1·0	Low-medium	Satisfactory in the upper range
0·75–1·3	1·0–1·75	Medium to high	Desirable range, no fertiliser needed
1·3–2·0	1·75–2·75	High to very high	Do not add fertiliser or allow compost to dry out. May require leaching

method, are usually present in only relatively small amounts. It is, however, possible for the compost to contain a large amount of urea which is a non-electrolyte and the true salinity or osmotic pressure would then be much higher than that indicated by a specific conductivity reading. Under these conditions, the true osmotic pressure could be measured only by a determination of the freezing point depression. The comparative osmotic pressures developed by a number of commonly used fertilisers have been included in Table 6.5.

Table 4.6. *Interpretation of the conductivity of a calcium sulphate extract for a standard soil of bulk density $1\,g\,cm^{-3}$ (ADAS System).*

Conductivity micromhos	Index	Interpretation
2 210–2 400	1	Glasshouse soils with low level of nutrients
2 410–2 600	2	Normal level of salt; for lettuce and flower production
2 610–2 700	3	Safe level for established flower crops and for tomato production
2 710–2 800	4	More sensitive crops may be restricted (lettuce, young cuttings and seedlings), safe for established carnations and tomatoes
2 810–3 000	5	Even less sensitive plants affected especially at young stage of growth
3 010–3 300	6	Growth restriction possible even with tolerant crops
3 310–3 700	7	Severe growth restriction likely even on tolerant crops, possible root damage
>3 710	>8	Excessively high concentration

Chapter 5

Nitrogen

Whilst from the viewpoint of the plant physiologist, all of the *essential* elements can be regarded as being of equal importance in the nutrition of the plant, in terms of the practical management of loamless composts nitrogen is probably the most important single element. It forms between 2 and 4% of the dry weight of the average pot plant and occurs in proteins, amino acids and enzymes. It is also an essential component of chlorophyll, the green pigment present in leaves which enables plants to utilise the sun's energy to combine carbon dioxide from the atmosphere with water vapour to form carbohydrates. This process is known as photosynthesis and provides man with his primary source of food. When nitrogen is deficient there is a marked reduction in the growth rate. Plants are tall and spindly, the upper leaves being narrow and erect whilst the lower leaves turn first pale green and then yellow as the nitrogen is withdrawn and transported to the actively growing regions. Often the older leaves produce bright colours before they die. When the nitrogen level in the compost is high in relation to the phosphorus and potassium, plants tend to make soft, vegetative growth and be less reproductive with delayed flowering. This situation is intensified when the plants are grown under conditions of low light intensity, high humidity and low water stress. Soft vegetative plants are more susceptible to fungal diseases such as botrytis. If the nitrogen level in the compost is further increased, there will be a corresponding increase in the soluble salt level, and eventually this will have the effect of producing hard, stunted plants.

Although the atmosphere contains approximately 80% of gaseous nitrogen, this is not available to the plants, apart from those cases where there is a symbiotic association with the nitrogen-fixing bacteria present in the soil. Excluding the few cases where nitrogen compounds are applied to plants as foliar sprays, notably the pineapple, virtually all of the nitrogen required must be absorbed through the roots in an inorganic form, predominantly as ammonium (NH_4^+) or nitrate (NO_3^-).

5.1 NITROGEN AND POT PLANTS

There are three principal reasons why a continuous supply of mineral nitrogen is of relatively greater importance when growing pot plants in loamless composts than it is for border and field grown crops.

Firstly, most of the materials used to make loamless composts either do not contain any significant amounts of nitrogen or, as in the case of peats, which may have from 1 to 3% of their dry weight as organic nitrogen, this is in a form which is not readily available during the life span of the average pot plant. This situation is shown in Figure 5.1 where the amount of nitrogen in the tissue of tomato plants grown in a peat-sand compost is compared with plants grown in a mineral soil compost. Both composts were supplied with adequate amounts of all nutrients except nitrogen which was omitted. Plants grown in the peat–sand compost had been able to absorb only 20% as much nitrogen as plants grown in the loam compost.

Secondly, when plants are grown in containers, their roots are restricted to a relatively small volume and, unlike border grown plants, they cannot

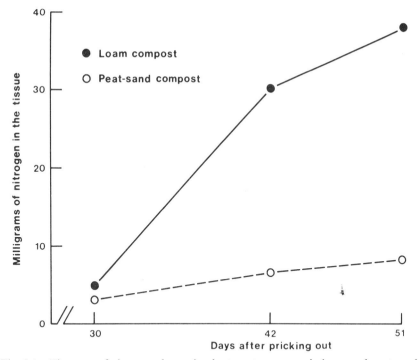

Fig. 5.1 The rates of nitrogen absorption by tomatoes grown in loam and peat-sand composts. Plant nutrients, except nitrogen, had been added in the base fertiliser. Forty-two days after pricking-out plants grown in the peat-sand compost had been able to absorb only one quarter of the nitrogen of plants grown in the loam compost.

84   Nitrogen

rely upon an expanding root system to provide a continuing supply of nitrogen. Once the roots have filled the volume of compost available and the plant is 'pot-bound', the mineral nitrogen will soon become depleted and the plants will show signs of nitrogen deficiency unless liquid feeding is given.

Thirdly, because of the small volume of the pot, the amount of water available to the roots is limited, and most pot-grown plants will require frequent watering, possibly three times per day under very bright summer conditions. It is virtually impossible to irrigate pot plants without causing some loss of nutrients by leaching, and as seen in Chapter 4 both nitrate and ammonium nitrogen are readily leached from most loamless composts. The loss of ammonium from the compost can however be reduced by the inclusion of some vermiculite.

Table 5.1. *The range of dry weights and nitrogen contents of some pot and bedding plants.*

Plant	Total dry weight (g)	%N	Total N (mg)	Growing period
Saintpaulia	3·2	2·1	67	15 weeks
Exacum	3·6	4·0	144	16 weeks
Tomato	4·1	3·7	152	10 weeks (Nov.–Jan. propagated)
Box of bedding begonias (32 plants/4·5 litres of compost)	16·2	3·7	599	8 weeks (summer)
Zinnia (ditto)	25·4	1·4	355	8 weeks (summer)
Chrysanthemum				
Winter	20·0	5·5	1 030	10 weeks
Summer	57·0	3·3	1 870	10 weeks
Cyclamen	26·6	3·9	1 037	1 year
Ficus	66·4	1·6	1 062	1 year

The total amount of nitrogen and the rate at which it is required by the plant depend largely upon the vigour of the species, and the manner in which the plants are grown. Temperature, light, water availability and size of container are some of the factors which control the rate and amount of growth. Some typical plant weights with the amount of nitrogen in their tissue, and the period over which they were grown, are given in Table 5.1. The contrast between the rates at which nitrogen was absorbed by pot chrysanthemums grown in summer, a crop which has a very high nitrogen requirement, and by the relatively slower growing cyclamen, is shown in Figure 5.2; both crops were grown in containers holding one litre of compost. During the last eight

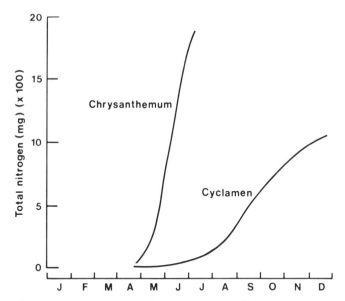

Fig. 5.2 Contrasting the rates and the total amount of nitrogen taken up by a pot chrysanthemum, grown in ten weeks in summer, and a cyclamen, grown over one year. Both plants were grown in one litre of compost and were given liquid feeding.

weeks of its ten-week growing period, the pot chrysanthemum absorbed nitrogen at the rate of 30 mg per day, the total nitrogen uptake being 1 870 mg. With the cyclamen, the maximum rate of nitrogen absorption between August and December was only 7 mg per day, and the total nitrogen uptake over the one year growing period was 1 030 mg. In its ten-week growing period the chrysanthemum had assimilated twice as much nitrogen as the cyclamen had during a whole year.

The maximum amount of mineral nitrogen that can safely be added in the base fertiliser varies with the plant species, but it is only of the order of 200 to 250 mg/l of compost. Quantities in excess of this can cause either a specific toxicity or a high salinity hazard. Nitrate nitrogen causes a big increase in the compost salinity, and whilst ammonium produced from urea does not ionise to any appreciable extent, and so is not detected by salinity measurements based on electrical conductivity, this nitrogen does, in fact, contribute to the osmotic pressure or the total salt stress experienced by the plant. Therefore only a small part of the total amount of nitrogen required by the average pot grown plant can be supplied as mineral nitrogen in the base fertiliser. Either the greater part of the nitrogen must be supplied in a slow release form, which gives a controlled release over the growing period, or the relatively small amount of mineral nitrogen that can safely be applied in the base fertiliser must be supplemented by frequent liquid feeding. This latter system of nutrition is discussed in detail in Chapter 9.

5.2 FORMS OF MINERAL NITROGEN

AMMONIUM AND NITRATE NITROGEN

Plants absorb only relatively small quantities of the various organic forms of nitrogen and most of the mineral nitrogen is taken up in the ammonium and nitrate forms. Whilst, frequently, no clear preference is shown by plants for either of these forms of nitrogen, there are certain exceptions, e.g. the inhibition of flowering in *Lemna perpusilla* at low concentrations of the ammonium ion but not by nitrate ions (Hillman and Posner, 1971). Often the apparent preference for a particular form of nitrogen is related to the accompanying effects or to other elements present in the fertiliser. For example, fertilisers such as ammonium sulphate and ammonium chloride, which do not have a metallic cation, are acidic and cause a reduction in the pH of the compost, whereas calcium nitrate and potassium nitrate add bases to the compost. Absorption of ammonium by the plant tends to lower the pH of the compost, whilst the uptake of nitrate nitrogen tends to raise the compost pH. Nitrogenous fertilisers also differ in the amount of salinity they produce in the compost per unit of nitrogen applied, see Table 6.5 (p. 127).

Under winter conditions, when the rate of photosynthesis is low, it is often safer to apply nitrate rather than ammonium nitrogen. Ammonium nitrogen combines rapidly with the carbohydrates in the roots and can lead to a condition of carbohydrate depletion, whereas nitrate nitrogen must be reduced to ammonium within the plant before it is fully assimilated. Young seedlings, low in carbohydrates, will often show signs of ammonium toxicity if the ammonium level in the compost is too high. Ammonium also depresses the uptake of other cations such as potassium and calcium. For example, tomatoes given ammonium-producing fertilisers are more prone to the physiological disease 'blossom-end rot', which is associated with a low calcium status, than are plants receiving nitrate nitrogen. Ammonium toxicity in plants has been studied by Maynard and Barker (1969). They found that bean, sweet corn, cucumber and pea plants grown in nutrient solutions were susceptible to ammonium toxicity; this was eliminated when the pH of the nutrient solution was maintained near to neutrality by the addition of calcium carbonate. Onion plants, however, were found to have a marked resistance to ammonium toxicity. Nelson and Hsiek (1971) investigated the toxicity of ammonium to pot-grown chrysanthemums in relation to the ammonium–potassium ratio in the leaves. Ammonium toxicity symptoms included a reduction in the growth rate, the development of small necrotic spots on the leaves, the production of thickened leathery leaves and the eventual death of the plant. They found that when the NH_4/K ratio in the leaves was less than

0·022, i.e. less than 2·2 units of ammonium per 100 units of potassium, there were no symptoms, but at ratios above 0·026 toxicity symptoms appeared. One of the incipient symptoms of ammonium toxicity in young tomato plants is the epinastic curving of the leaves and the premature death of the cotyledons. These symptoms are especially noticeable under the low light conditions of winter.

Kohl et al. (1955) found that ammonium nitrogen in moderate excess was much more harmful, and restricted the growth of Saintpaulia more than did nitrate nitrogen. Jungk (1968) compared the effect of ammonium and nitrate nutrition on *Sinapis alba* grown in sphagnum peat at pH 6·6. At rates above 100 ppm of nitrogen the plants receiving ammonium made much less dry weight and produced fewer flowers than plants receiving nitrate nitrogen. Gasser (1964) found that ammonium salts were more damaging than potassium salts to germinating seeds of kale, barley and wheat, suggesting a toxic effect of ammonium.

Ericaceous plants in general, and azalea in particular, are, however, good examples of plants which prefer ammonium to nitrate nitrogen. Colgrave and Roberts (1956) found that azalea plants receiving NH_4—N gave better growth and produced less chlorosis than plants having NO_3—N. The chlorosis was found to be associated with the pH of the leaf tissue; ammonium reduced the uptake of the other cations and thereby produced a lower pH. Nitrates, however, increased the absorption of bases, and, as a consequence of the higher pH, inactivated the iron in the leaves. In experiments made at the Glasshouse Crops Research Institute to investigate the response of plants to the form of nitrogen in the liquid fertiliser, chrysanthemums grown in peat–sand composts were given liquid feeds in which the nitrogen was either all in the form of nitrates, all as ammonium or partly in each form. Plants receiving only nitrate-nitrogen had paler leaves with less chlorophyll than plants receiving some or all of the nitrogen in the ammonium form; this difference in leaf colour was not always directly related to the final pH of the compost. North and Wallace (1959) found that macadamia plants grown with only nitrate-nitrogen were liable to show leaf chlorosis and even some necrosis; they were also found to be low in iron. Plants receiving some of their nitrogen in the ammonium form did not show any of these symptoms.

Two hazards, however, associated with the use of ammonium-producing fertilisers are free ammonia (NH_3) and nitrites.

Free ammonia. In mineral soils, the ammonium ions, which can be derived from either an ammonium fertiliser, such as ammonium sulphate, or from the mineralisation of organic fertilisers, such as hoof and horn and urea-formaldehyde, are rapidly converted into nitrites and then into nitrates. In composts, this process is sometimes restricted if the population of the nitrifying bacteria is low, or if they are temporarily eliminated from the compost

88 *Nitrogen*

following heat sterilisation, thereby causing an accumulation of either ammonium or nitrite-nitrogen. A simplified diagram showing the nitrogen transformations which occur in soils is given in Figure 5.3. In soils containing ammonium ions, free ammonia can occur when the pH is just below the neutral point; the concentration of free ammonia and its loss from the compost then increases very rapidly as the pH rises. At high pH values, the non-ionised ammonia is freely able to enter the root cells where it quickly reaches toxic levels. The concentrations at which free ammonia affected the growth of sudan grass and cotton seedlings were investigated by Bennett and Adams (1970); they concluded that the initial concentration for incipient toxicity was 0·15 to 0·20 m Molar of non-ionised ammonia in the soil solution.

A practical example of the effect which different sources of ammonium fertilisers can have on plant growth is shown in Figure 5.4. Tomato seedlings were grown in winter in peat–sand composts having a pH of 6·0 (M/100 $CaCl_2$) before the various nitrogenous fertilisers had been added. One week after the fertilisers had been mixed into the composts, the pH values were:

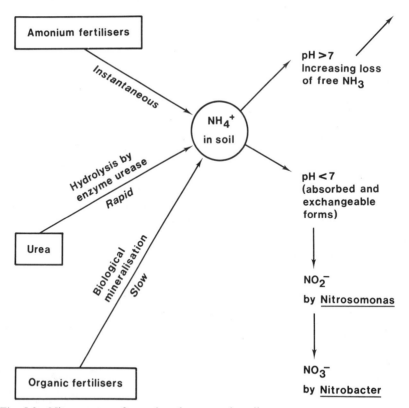

Fig. 5.3 Nitrogen transformation that occur in soils.

compost with ammonium carbonate pH 6·50, compost with urea pH 6·54, compost with ammonium sulphate pH 6·08. Seven weeks after pricking-out, the mean fresh weights of the plants were: ammonium carbonate 0·49 g, urea 1·18 g, ammonium sulphate 5·54 g. Plants grown with calcium nitrate as the nitrogen source (not shown in the illustration) had a mean weight of 8·49 g.

Fig. 5.4 Tomatoes grown in winter in composts with equal amounts of ammoniacal nitrogen from three sources.
Left: ammonium carbonate; *centre:* urea; *right:* ammonium sulphate.

Nitrite. Another toxin associated with ammonia and ammonia-producing fertilisers is nitrite. Normally ammonium is quickly converted to nitrate-nitrogen by the nitrifying bacteria, *nitrosomonas* and *nitrobacter*. When soils have been steam sterilised or when large concentrations of ammonium have occurred, the *nitrobacter* can be inhibited for a period, resulting in a build-up of nitrites to toxic levels. High levels of nitrite-nitrogen frequently occur when urea has been used in the base fertiliser. Bunt and Adams (1966, *b*) found the nitrite-nitrogen level in a peat–sand compost made with urea, rose to 147 mg/l; this was equivalent to 50% of the mineral nitrogen in the compost. When hoof and horn was used as the source of base fertiliser nitrogen, the maximum recorded level of nitrite-nitrogen did not exceed 8% of the mineral nitrogen level. Whilst nitrites accumulate more rapidly under alkaline conditions, it is generally agreed that nitrite is more toxic to plants under acid conditions. Toxicities and reductions in growth have been reported at concentration around 10 mg/l.

With field grown crops, the greater part of the nitrogen absorbed by the plant will be in the nitrate form; any ammonium supplied in relatively infrequent fertiliser dressings is rapidly converted into nitrates. With pot grown plants, however, a much greater proportion of the nitrogen is taken up as ammonium. This is principally due to:

(a) steam sterilisation causing a temporary delay in the production of nitrates, or
(b) the naturally slower rate of nitrification in peat composts, and
(c) the production of ammonium from slow-release forms of fertilisers or
(d) the use of liquid fertilisers containing ammonium. Some crops are given liquid fertilisers as frequently as twice per day.

MINERAL NITROGENOUS FERTILISERS

Whilst there are many mineral nitrogenous fertilisers, in practice only a few are in general use in base fertilisers for potting composts.

Ammonium sulphate [$(NH_4)_2SO_4$] contains 21%N, all in the ammonium form. In common with most fertilisers which supply ammonium, this fertiliser is acid-forming and its use results in a fall in the compost pH. Under agricultural conditions, it is generally considered necessary to apply an equal weight of calcium carbonate in order to maintain the pH level.

Calcium nitrate [$Ca(NO_3)_2$] contains 17·1%N and 24%Ca; this fertiliser will increase the pH of the compost.

Ammonium nitrate (NH_4NO_3) 35%N, supplies equal amounts of ammonium and nitrate nitrogen. In its earlier form, this fertiliser was very deliquescent and had to be stored under dry conditions. To improve its handling and storage properties it is now sold in 'prilled' form. During manufacture, the concentrated ammonium nitrate solution, which has about 2% of magnesium nitrate added as an internal desiccant, is sprayed to form small droplets which are cooled and solidified by convection into prills. Whilst ammonium nitrate is classified as an oxidising agent, it will not burn on its own; only if there is a fire within a confined space will ammonium nitrate increase the fire and explosive risk. *Under normal conditions of usage this fertiliser is quite safe and does not present any risk.*

Urea [$CO(NH_2)_2$] contains 46%N. Whilst from its formulation it can be considered as being an organic source of nitrogen, in practice it behaves as an inorganic fertiliser, being hydrolysed in the compost within a few hours to ammonium carbonate by the enzyme urease. Urea should not contain more than 2·5% of biuret, which at high concentrations is toxic to plants. Hazards associated with the use of this fertiliser are free ammonia and nitrites.

Calcium-ammonium nitrate. This fertiliser is not available in Britain but is used in both Norway and Ireland as a nitrogen source for peat based composts. It is made from ammonium nitrate and calcium carbonate. The analysis and the name of the fertiliser can vary with the country of origin. Nitrochalk contains 15·5%N, whilst the name calcium-ammonium nitrate is reserved for mixtures containing 21 or 26%N. Cal Nitro and Nitrolime are names used for lower grade materials. If it is intended to use these fertilisers in loamless composts, it is essential that both the nitrogen and calcium contents are first checked.

Potassium nitrate (KNO_3) contains 13%N and 38%K. The high potassium content means that it is primarily regarded as being a potassium fertiliser.

With all these fertilisers the nitrogen is in an immediately available form; this has the advantage that the amount of mineral nitrogen present in the compost is under direct control and can be regulated according to the plant's requirements. There is also the disadvantage of increasing the compost salinity together with the greater risk of loss by leaching. To avoid these disadvantages and to allow more nitrogen to be added in the base fertiliser, it is common practice to use some 'slow-release' forms of nitrogen.

5.3 SLOW RELEASE FERTILISERS

This term is used to describe fertilisers which release nutrients slowly to plants over an extended period. These fertilisers can be classified into three groups depending upon their composition and mode of release, viz. (a) organic, (b) compounds having slow rates of mineralisation of dissolution, (c) coated fertilisers.

ORGANIC

With traditional organic fertilisers, such as hoof and horn (13%N) and dried blood (10–13%N), mineralisation of the organic nitrogen is by fungal and bacterial decomposition. The various steps in the mineralisation process can be represented by the following simplified stages:

Amino acid production

$$\text{Organic nitrogen in fertilisers and protein nitrogen in organic matter} \xrightarrow{\text{By enzymes}} \text{Free amino acid nitrogen} (NH_2) + CO_2 + \text{Energy} \quad (I)$$

Ammonification

$$R-NH_2 + HOH \rightarrow NH_3 + R-OH + \text{Energy} \quad (II)$$

(R can be a hydrogen atom, short carbon chain or methyl group)
Stages (I) and (II) are preformed by heterotrophic micro-organisms that rely

on organic substances for their energy. These organisms are unable to utilise carbon dioxide and inorganic compounds.

The ammonia (NH_3) produced in stage (II) combines with carbonic acid and other soil acids to form the ammonium (NH_4^+) ion. Only under neutral or alkaline conditions will significant amounts of ammonia persist as free ammonia.

Nitrification

$$2NH_4^+ + 3O_2 \rightarrow 2NO_2^- + 2H_2O + 4H^+ + \text{Energy} \quad \text{(III)}$$

$$2NO_2^- + O_2 \rightarrow 2NO_3^- + \text{Energy} \quad \text{(IV)}$$

Stage (III) is preformed by *nitrosomonas* and stage (IV) by *nitrobacter*. Both of these bacteria are *chemoautotrophs*, i.e. they obtain their energy for

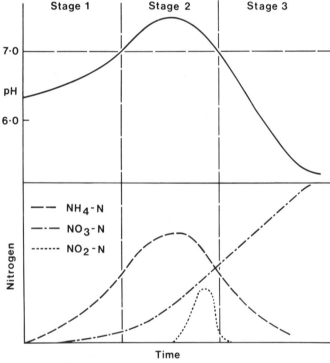

Fig. 5.5 Diagrammatic representation of the changes which occur in the forms and amounts of mineral nitrogen, and in the compost pH, when organic type fertilisers are used in peat-sand composts. *Stage 1:* ammonium nitrogen produced, pH may rise to the neutral point. *Stage 2:* more ammonium produced, pH rises above the neutral point and free ammonia present; nitrites may also be present; nitrates start to form. *Stage 3:* ammonium falls, nitrates increase and the pH is now below the original value.

growth from the oxidation of inorganic compounds and their carbon by assimilation of carbon dioxide. The nitrifying process requires molecular oxygen and will only proceed rapidly in well aerated soils or composts. It will also be seen that hydrogen ions (H^+) are released during the nitrification process, and the continued use of organic or ammonium fertilisers will lead to the acidification of the compost.

In peat based composts, the micro-organisms which first break down the organic nitrogen to ammonium are more active than those which convert ammonium to nitrite and then to nitrate. This leads to a build-up of ammonium which under certain conditions can cause the compost pH to rise by as much as one whole pH unit. If the pH exceeds the neutral point there is the risk of free ammonia toxicity and of micro-element deficiencies, primarily of boron. This situation can persist until the pH drops with the conversion of the ammonium to nitrates. The related changes in the forms of nitrogen and the compost pH, which occur during the mineralisation of organic nitrogen, are shown in Figure 5.5. Factors which increase the rate of mineralisation of hoof and horn are a reduction in its particle size, an increase in the temperature of the compost up to 40°C and an increase in the water content until it approaches the 'container capacity' level. Under glasshouse conditions of high compost temperature and available water supply, about 70% of the nitrogen can be mineralised within 30 days. Because of their relatively quick rates of release under these conditions and the high risk of toxicities, these fertilisers are not as widely used in loamless composts as they were in the mineral soil-based John Innes composts.

COMPOUNDS WITH LOW RATES OF MINERALISATION OR DISSOLUTION

Urea-formaldehyde (UF). Originally developed by workers in the United States Department of Agriculture, this fertiliser is formed as a controlled condensation product of urea and formaldehyde. For fertiliser purposes, the molar ratio of urea to formaldehyde is between 1·2 and 1·4 and the product has at least 35% nitrogen; most of the commercial UF fertilisers contain 38%N. Long and Winsor (1960) investigated the rates at which different urea-formaldehyde compounds were mineralised in soil, and found that methylene-diurea and dimethylene-triurea both mineralised too rapidly to be of value as slow release nitrogen sources, whereas tetramethylene-pentaurea has too slow a rate of decomposition. Trimethylene-tetraurea had the most useful mineralisation rate of all the compounds tested. Commercial UF fertilisers consist of mixtures of these compounds with some free urea also present and it is customary to state their 'availability index' (AI) as a means of assessing their probable rates of mineralisation in soils:

$$AI = \frac{\% \text{ cold water insoluble N} - \% \text{ hot water insoluble N}}{\% \text{ coldwater insoluble N}} \times 100$$

94 *Nitrogen*

This index approximates to the amount of insoluble nitrogen which nitrifies in an average soil in about six months; for use in potting composts the fertiliser should have an AI value of between 40 and 45.

Urea-formaldehyde is mineralised in the compost by bacterial action, with some chemical hydrolysis also occurring in the initial stages. The rate of mineralisation is not significantly affected by the particle size of the fertiliser but is increased by high compost temperatures and by low pH values (Winsor and Long, 1956). This effect of the compost pH on the mineralisation rate is in contrast to that found with hoof and horn; comparison of the mineralisation rates of these two fertilisers in peat–sand composts at two pH values is shown in Figure 5.6. It has been estimated that approximately one third of the

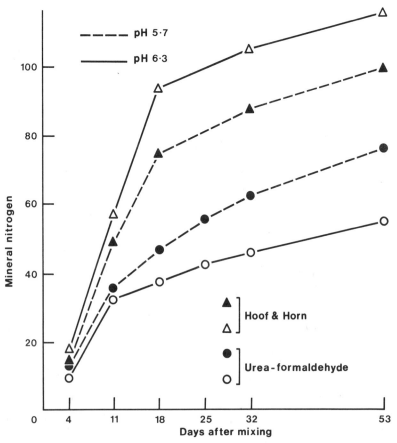

Fig. 5.6 Contrasting the effect of the compost pH on the mineralisation rates of hoof and horn and urea formaldehyde. With hoof and horn the high pH gave the fastest rate of mineralisation, whereas with urea formaldehyde the fastest rate of mineralisation occurred at the low pH.

nitrogen present in urea-formaldehyde is available to plants within a few weeks, a further third within several months and the remaining third within 1–2 years. For this reason, it is unlikely that the average pot-grown plant recovers more than 50% of the applied nitrogen. With border grown crops, however, some of the remaining nitrogen would be recovered by successive crops. Commercial urea-formaldehyde fertilisers available in Britain are 'Nitroform' and 'Ureaform'.

Crotonylidene-diurea (CDU). German workers have developed a fertiliser based upon the condensation of urea with crotonaldehyde. This crotonylidene-diurea has the trade name 'Floranid' and contains 28%N, 90% of which is in a slow-release form and the remaining 10% is present as nitrate-nitrogen. Nitrogen release is by bacterial and fungal breakdown and, as with urea-formaldehyde, the decomposition rate is inversely related to the compost pH. Very little nitrogen is released during the first six weeks if the compost pH is about 6·0. The rate of mineralisation is also dependent upon the soil temperature: at the low soil temperatures of 7–10°C virtually no nitrogen is released. A reduction in the particle size of the fertiliser also gives a higher rate of mineralisation. Results of incubation tests in peat–sand composts at the Glasshouse Crops Research Institute have shown this material to have the slowest release rate of all the slow-release fertilisers tested. It is widely used in peat composts in Germany but is not at present available in Britain where urea-formaldehyde is the most widely used slow-release source of nitrogen.

Isobutyridene-diurea (IBDU). This material is a condensation product of urea and isobutyraldehyde in a 2:1 mole ratio; it has 32%N and a very low solubility in water. Nitrogen release is by dissolution and hydrolysis and is not significantly affected by microbial activity. The principal factor known to increase the rate of hydrolysis is a decrease in the particle size; the normal size of material used in composts is 0·7 to 2·0 mm. An increase in soil temperature has only a relatively small effect on the release rate and plants are able to use the fertiliser most effectively at pH values between 5·5 and 6·5. It can normally be safely used at high rates of application but a mild toxicity and some depression in growth has been observed when used at the rate of 500 mg N/l of compost, equivalent to 1·56 g of fertiliser per litre.

IBDU is distinct from UF and CDU in that the release mechanism is chemical rather than microbiological.

Magnesium–ammonium–phosphate. Several of the divalent metals form compounds with ammonium phosphate which have a low rate of solubility. Because of the risk of toxicities to plants, magnesium is preferred to either the manganese, zinc, copper or molybdenum–ammonium phosphates; magnesium ammonium phosphate has an analysis of 8%N, 17·4%P 14% Mg.

A low compost pH and high moisture content increase the rate of release; temperature has only a very slight effect on the rate of dissolution. The principal method of controlling the release is by regulating the particle size; small particles have a much higher rate of release because of their higher surface: volume ratio. Release rates have been found to be greater than would be expected purely from the water solubility values, and it is concluded that diffusion of the ammonium away from the particle surface and also the conversion of the ammonium to nitrate is responsible for the higher rate of dissolution. The grade of material suitable for use in composts has 80% of the particles in the 1·5 to 3·0 mm range; larger particles give less uniformity of fertiliser distribution. This material is available in the U.S.A. under the trade name 'Magamp'. In addition to the compound having the 8–17–0–14 analysis, another material has some potassium ammonium phosphate included, giving an analysis of 7% N, 17·4%P 5% K, 13% Mg. In Britain a product known as 'Enmag' contains 5% N, 10·5%P 8·3%K, 10% Mg; the potassium is present as potassium sulphate.

Two effects sometimes associated with magnesium ammonium phosphates have been potassium deficiency induced by the high rates of magnesium, and an iron deficiency induced by the high rates of phosphorus forming insoluble iron phosphates. Potassium deficiency is controlled by increasing the amount of potassium in the base fertiliser, and the iron deficiency can be controlled by applying a 0·1% solution of iron chelate as required. The value of these materials as slow release sources of nitrogen has generally been over emphasised; their low nitrogen and high phosphorus content does not commend them for this purpose. The high rate at which the water-soluble phosphatic fertilisers are lost from peat–sand composts by leaching rather suggests that the greatest potential use for the metal ammonium phosphates is as slow-release sources of phosphorus.

Oxamide. Oxamide is a diamide of oxalic acid with 31·8%N. Its dissolution rate is controlled by the particle size; large particles give a slower initial release rate, but this is sustained for a longer time. The recovery of nitrogen by plants has been highest in soils with high pH values. Whilst this material has given good results in experiments, it is not generally available owing to its high cost of production.

COATED FERTILISERS

Various materials have been used for coating inorganic fertilisers to reduce their immediate availability to plants and so give a slow-release action. The method by which the fertiliser is released from the coating varies. With some materials, water vapour diffuses through the coating and eventually causes it to rupture; with other materials, the dissolved fertiliser diffuses out through very small pores in the coating, whilst in other cases the fertiliser is released through the slow microbial decomposition of the coating. Materials most widely used for coating the fertilisers are acrylic and epoxy resins, and sulphur.

Resinous coated fertilisers. Nutrient release is by diffusion and is controlled by the thickness of the coating which usually varies between 8 and 15% of the total weight of the coated fertiliser. Soil temperature also helps in controlling the rate of release; at a temperature of 20°C the release rate was found to be double that at 10°C (Oertli and Lunt, 1962). Neither the pH of the soil nor its degree of microbial activity was found to have any significant effect upon the release rate. This type of fertiliser is available either as a straight nitrogen source, usually as coated urea, or as a complete N P K fertiliser. In the latter case, the nitrogen, which is present in approximately equal amounts of ammonium and nitrate, is released more rapidly than either the phosphorus or the potassium. Two proprietary fertilisers of this type, sold under the name 'Osmocote', have an analysis of 14% N, 6·1% P 11·6% K (14:14:14) and 18% N, 2·6% P 10% K (18:6:12).

Sulphur coated urea. This type of fertiliser has been developed in the USA by the Tennessee Valley Authority and by Imperial Chemical Industries in Britain. The USA product, known as SCU, has 38·1% N with the coating forming 20% of the total weight, 16% being sulphur, 3% wax sealant and 1% a conditioner; there is also a microbicide incorporated, to protect the sealant against rapid microbial decomposition. Sulphur coated urea can be prepared with different dissolution rates. Furuta, Sciaroni and Breece (1967) used materials having dissolution rates of 6, 5 or 1% per day, and found the two materials with the highest rates of dissolution to be toxic to some ornamentals. They recommended the material with the 1% dissolution rate be used at 100 g N/cu.m of compost (0·26 kg/cu.m of fertiliser). The dissolution rate measures the rate at which the nitrogen is released under standard laboratory conditions and is defined as:

$$\text{dissolution rate} = \frac{\text{wt of fertiliser dissolved in water at 38°C in 24 hrs} \times 100}{\text{original weight of fertiliser}}$$

The British product is known as 'Gold-N' and contains 32% N. As with the USA product, it consists of urea which is coated with sulphur and a sealant, but it does not contain a microbicide. No direct comparisons of the dissolution rates of the two products appears to have been made, but Prasad and Gallagher (1972) concluded that the USA material mineralised too slowly in peat to be a suitable source of nitrogen for tomato nutrition; in 8 weeks the sulphur-coated urea had released only slightly more mineral nitrogen than the urea-formaldehyde. Working with the British product 'Gold-N', Bunt (1974, *a*) found this had a significantly faster release rate than urea-formaldehyde, comparable values of mineral nitrogen after 4 weeks being 200 and 126 mg N/l respectively. This suggests that the USA product has a slower dissolution rate than 'Gold-N'.

Results of studies at the Glasshouse Crops Research Institute on the mineralisation and dissolution rates of several of these slow-release fertilisers in peat–sand composts are given in Table 5.2.

Table 5.2. *Nitrogen mineralisation rates of several slow-release fertilisers in peat-sand substrates (mg water-soluble nitrogen per l).*

Fertiliser	Time (days)			Mineral N present after 28 days incubation relative to hoof and horn
	7	14	28	
Hoof and horn	94	137	154	100·0
Floranid	62	78	94	52·0
Urea formaldehyde	40	86	126	84·7
IBDU	50	85	133	86·5
Gold-N	98	154	200	129·9
Enmag	104	135	151	97·9
Mag Amp	105	138	158	102·7
Plantosan 4D	102	149	172	111·5
Plantosan SR	116	153	189	123·0
Osmocote (14:14:14)	90	174	223	145·2

Although the effect of steam-sterilising the composts after using slow-release fertilisers has not been adequately investigated, in general it appears that the steam sterilisation of composts containing coated fertilisers or urea formaldehyde has resulted in an increased rate of nitrogen release and a rapid build up in salinity. For this reason, steam sterilisation of the compost after applying slow-release fertilisers is not recommended. Magnesium–ammonium–phosphate, however, is apparently not affected by steam sterilisation.

5.4 CHOICE OF FERTILISER TYPE

Several factors must be considered in deciding which of these two types of fertiliser to use in preparing loamless composts, i.e. the inorganic fertilisers or the slow-release types. The most important points for consideration will be:

(a) the probable length of the period between mixing and using, i.e. the storage period,
(b) the type of crop and season,
(c) the ability to give liquid fertilisers and the relative convenience of the two systems,
(d) the relative costs of the various fertilisers.

STORAGE OF MIXED COMPOST

When determining the type of fertiliser to use, the storage period of the compost is an important factor to consider. It has already been stated that whilst ammonification of the organic type of fertilisers in peat based composts is fairly rapid, nitrification is much slower. If composts made with this type of fertiliser are stored for a period of more than one week before they are used, there is a distinct possibility of toxicities developing, principally from the build-up of ammonium and the accompanying rise in pH.

This situation is clearly seen from the work of Bunt and Adams (unpublished), where the interaction of the storage period and the source of nitrogen was investigated by preparing two composts, one with hoof and horn, and the other with calcium nitrate. After mixing, the two composts were stored, and samples used over nine successive weeks to grow tomato seedlings whose growth rates were recorded from weekly measurements of the leaf lengths. The effect which the length of the mixed compost's storage period had on plant growth has been summarised in Figure 5.7. The total length of the leaves of plants in the hoof and horn compost after an eight-week growth period has been expressed as a percentage of the comparable plants grown in the calcium nitrate compost. Also included in Figure 5.7 are the ammonium, nitrate and pH levels of the stored hoof and horn compost at the time it was used to prick out the tomato seedlings. There was virtually no change in the stored calcium nitrate compost.

When the compost was used without any storage period, plants in the hoof and horn compost were only very slightly inferior to those grown in the calcium nitrate treatment. As the storage period was increased, there was a progressive decrease in the growth of the plants in the hoof and horn compost. After three weeks of storage, plants in the hoof and horn compost made only 10% of the growth of plants in the corresponding calcium nitrate compost. As the storage period was further increased, the growth of plants in the hoof and horn composts steadily improved in relation to the corresponding calcium nitrate composts, until at six weeks after mixing growth was essentially similar in both composts. It will be seen that, at this time, the ammonium and pH levels in the stored compost were falling and nitrates were increasing. The effect which the length of the storage period has on plant growth will of course depend upon several factors, i.e. the plant species, the season, the temperature and moisture content of the compost, the particular form of fertiliser used and its rate of application. Results shown in Figure 5.7 serve only to illustrate a principle rather than to define the periods when the compost is either 'safe' or 'unsafe' to use. It is certainly safer to avoid using organic-type fertilisers, if there is any likelihood that the compost will not be used immediately after mixing, especially in the case of seedlings or young plants grown in winter.

If a compost containing too much slow-release nitrogen has inadvertently

100 *Nitrogen*

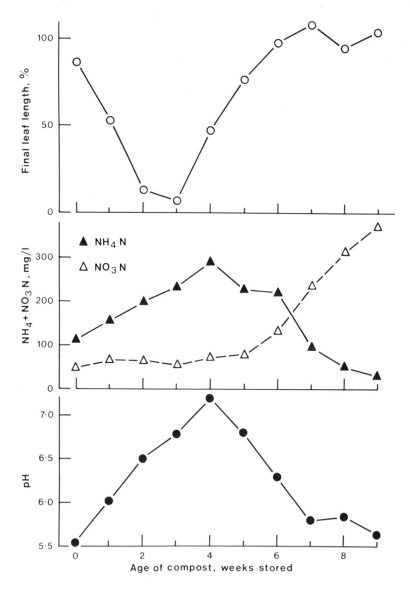

Fig. 5.7 Effect of the form of nitrogen and duration of compost storage on the growth of tomato in winter. *Top:* Growth of tomatoes in composts made with hoof and horn, and stored for varying periods, expressed as a percentage of the growth of plants in composts made with nitrate nitrogen and stored for similar periods. Growth was assessed by the total length of the leaves eight weeks after pricking-out the seedlings. *Centre:* the change in the ammonium and nitrate nitrogen levels of the hoof and horn composts during storage. *Bottom:* the change in the pH of the hoof and horn compost during storage.

been used, the nitrogen content cannot be reduced easily. Whilst both the NH_4^+ and NO_3^- forms of nitrogen can be readily leached from the compost, other nutrients will also be removed, and excessive watering of plants in non-porous containers in winter is better avoided. One approach to the problem of high levels of nitrogen in composts, caused by the continued release of nitrogen from slow release fertilisers in excess of the plants' requirements, has been to apply a solution of sucrose. This changes the carbon:nitrogen ratio in the compost, and some of the mineral nitrogen is temporarily immobilised. Solutions of 1 kg of sucrose per 30 litres of water (2 lb. per 6 gall.) have been applied to geraniums at the rate of 150 ml per 10-cm pot (6 fl.oz. per 4-in. pot) with beneficial results.

Free ammonia. Some of the toxicities associated with the storage of composts containing organic forms of nitrogen can be attributed, in part, to the production of free ammonia, and, in part, to the effect the increase in pH has on the availability of micro-elements, principally boron. In incubation experiments, the production of free ammonia has been found to parallel the rise and fall in the compost pH, but normally the loss of free ammonia has not been large in relation to the amount of nitrogen in the added organic fertiliser. Free ammonia readily penetrates cell membranes, however, and is therefore much more toxic to plants than the ammonium ion. In experiments at the Glasshouse Crops Research Institute, the maximum recorded loss of free ammonia occurring over a period of 42 days was 19·5 mg N/l of compost. A total of 850 mg of organic nitrogen had been applied, and this loss of free ammonia was equivalent to 2·3% of the added nitrogen or 3·8% of the final mineral nitrogen value (Fig. 5.8).

Micro-element availability. When composts are made with organic forms of nitrogen and then stored for a period before being used, the availability of the micro-elements is reduced because the build-up of ammonium causes a rise in the compost pH. The significance of this for the growth of plants in stored composts is shown in Figure 5.9. By comparison with plants grown in the freshly mixed compost, those grown in the stored compost had made very little growth. Plants growing in the compost to which micro-elements in the form of inorganic salts had been added before storage, however, made normal growth. Subsequent work has shown that the temporary unavailability of boron is one of the principal factors involved. Because of the risk of toxicity from boron salts, however, it is preferable to apply micro-elements in fritted form.

Some of the other factors to consider in making the choice between slow-release and mineral forms of fertiliser are less easy to quantify. For example, the ability to apply fertilisers in liquid form varies widely between nurseries. Some growers rely on using mineral fertilisers in the compost, supplemented

Nitrogen

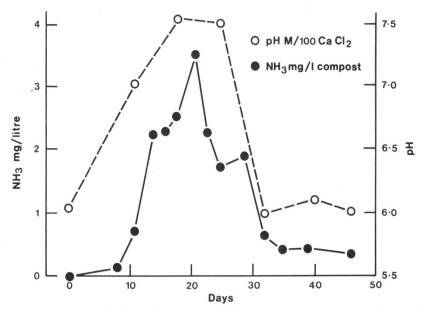

Fig. 5.8 The production of free ammonia from a compost made with organic nitrogen follows the change in the compost pH.

Fig. 5.9 Changes in the pH and forms of mineral nitrogen in stored composts, made with organic nitrogen, can also induce micro-element deficiencies, especially of boron. *Left:* antirrhinum seedling pricked out into freshly mixed compost, no micro-elements added. *Centre:* compost stored for three weeks after mixing, no micro-elements added. *Right:* micro-elements added during mixing, compost stored for three weeks before seedling pricked-out.

with liquid feeding, whilst other growers prefer slow-release fertilisers, because of the difficulty in obtaining sufficient skilled labour to operate liquid feeding equipment.

5.5 NITROGEN AND PEAT

Peats differ from many other materials used to make loamless composts because they contain relatively large amounts of nitrogen, usually between 1 and 2·5% of their dry weight, whereas perlite, plastics, sand and vermiculite, etc., contain little or no nitrogen. To put this apparently high nitrogen content of peat into perspective, however, the following three factors must be considered. Firstly, the nitrogen is largely in an organic form and is therefore not immediately available. Second, whilst peats may appear to contain 10 or more times as much total nitrogen as mineral soils, their bulk density is only about 1/10th or less of that of a mineral soil. When the amount of nitrogen is considered in relation to the *volume* rather than the *weight* of peat in which the plant is growing, there is a corresponding reduction. Third, peats contain some organic matter that is fairly readily decomposed and, during the decomposition process, some of the mineral nitrogen will be used by the micro-organisms, and will not, therefore, be available to the plants. This situation is known as 'nitrogen immobilisation'.

Often the carbon:nitrogen ratio is used as a means of categorising soils and other materials with regard to their probable behaviour in the mineralisation of organic nitrogen or the immobilisation of mineral nitrogen during cultivation.

CARBON:NITROGEN RATIO (C:N)

Quite wide differences exist between types of peat with respect to their C:N ratios: these vary from about 40:1 for some young sphagnum peats to 20:1 for sedge peats. By contrast, the mineral soils of Britain have values nearer to 10:1. However, too much reliance must not be placed upon the C:N ratio as giving a simple measure of the availability of nitrogen in various materials; much will depend upon the form of the organic matter. For example, Bollen (1953) found a very high C:N ratio of 729:1 for Western Red Cedar sawdust, but, because the material contained 60–70% of lignocellulose which decomposes very slowly, less nitrogen was required to offset the lock-up of nitrogen by the micro-organisms than would be indicated from the C:N ratio. In practice, it is found that the amount of carbon left after a period of decomposition is primarily determined by the chemical composition of the material, rather than by the C:N ratio; materials high in lignins decompose very slowly, whilst hemicellulose and cellulose decompose rapidly. Puustjärvi (1970, *a*) found the easily decomposable organic matter in various peats ranged from 65·6% for *Sphagnum fuscum* to 15·2% for a *Bryales-Carex* peat. The nitrogen requirements of *S. fuscum* peats will therefore be greater than that of sedge

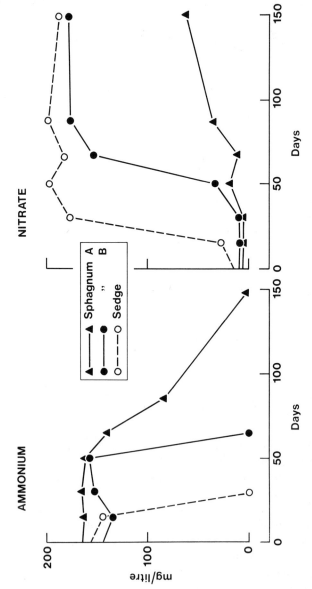

Fig. 5.10 Comparative rates of conversion of ammonium sulphate to nitrate nitrogen in three peats. Sedge peat showed the most rapid rate of nitrification, and sphagnum peat A the slowest rate of nitrification.

peats and, during the early stages of decomposition, an increased rate of nitrogen supply will be required to prevent the plants showing nitrogen deficiency symptoms. Later, when the rate of decomposition is reduced, some of this immobilised nitrogen will be released to the plant upon the death of the micro-organisms. This immobilisation of nitrogen and its subsequent release will have the greatest effect with border crops which are grown in relatively large volumes of compost, and where liquid feeding is not of such importance as it is with pot grown crops.

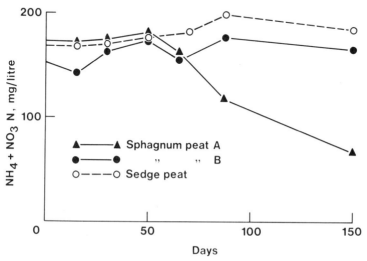

Fig. 5.11 Immobilisation of nitrogen in peats. In both the sedge peat and sphagnum peat B, there was no reduction in the amount of water soluble nitrogen, but in sphagnum peat A approximately 50% of the nitrogen was immobilised after 150 days.

Some results obtained at the Glasshouse Crops Research Institute, on nitrogen studies with different types of peat, are shown in Figures 5.10 and 5.11. Known amounts of ammonium sulphate, together with other base fertilisers supplying P, K, Ca, Mg and micro-elements, were added to three peats, i.e. sphagnums A and B, and also a sedge peat. The composts were moistened, placed in an incubator and ammonium and nitrate levels were determined at intervals. As seen in Figure 5.10, conversion of the ammonium into nitrate nitrogen was fairly rapid in the sedge peat compost, and slower in the sphagnum composts; sphagnum peat compost A being significantly slower than sphagnum peat B. Although none of the peats had been steam sterilised, the slow formation of nitrates in the sphagnum peats is similar to the delay found in mineral soils after steam sterilisation. Under glasshouse conditions, the time taken for the nitrifying bacteria to build up in population will be primarily dependent upon four factors: (a) the temperature of the compost, (b) its moisture content, (c) its pH and (d) the degree of bacterial inoculation. The most important sources of inoculation are the carry-over of inoculum on the roots during pricking-out or repotting from a compost in which the nitrifying bacteria have already built up a large population.

In Figure 5.11, the total amount of mineral nitrogen, i.e. the NH_4 plus NO_3 forms, present in the three peats during the course of the experiment shows that, after day 50, a significant amount of the mineral nitrogen in sphagnum peat A compost had been immobilised. No reduction in the mineral nitrogen levels had occurred in the composts made from either sphagnum peat B or the sedge peat. Results from this and other experiments show that usually the ammonium form of nitrogen is preferentially utilised by the bacteria in decomposing the peat.

Chapter 6

Other macro-elements

In addition to nitrogen, there are a number of other elements which are essential to plant growth; those required in relatively large amounts are known as *macro-* or *major* elements, whilst those required only in relatively small amounts are known as *micro-* or *trace* elements. Irrespective of the actual quantities required by the plants, however, both groups are equally *essential*; only a few elements such as barium, fluorine, iodine, and strontium do not appear to be essential for plant growth. The essential macro-elements are nitrogen, phosphorus, potassium, calcium, magnesium and sulphur.

Most of the materials described in Chapter 2 as being suitable for making loamless composts are deficient in several or all of the macro-elements, by comparison with a fertile mineral soil, two exceptions being peat, which has a relatively high organic nitrogen content, and vermiculite, which contains large amounts of potassium and magnesium. Relatively large quantities of fertilisers may therefore be required in order to raise the nutrients to suitable levels, if these materials are used to make potting composts.

The actual amount of fertilisers required to prepare potting composts will depend upon three principal factors:

(1) The amount of nutrients already present in the bulk materials and the rate at which they become available to the plant.
(2) The behaviour of the fertilisers when added to the compost; do they remain largely in a water-soluble form or are they liable to fixation and then become unavailable? Does the material have a significant cation exchange capacity?
(3) The species of plant to be grown; has it a short growing period and a high nutrient demand, e.g. pot chrysanthemum, or does it have a low demand for nutrients, e.g. calceolaria? Is there a known requirement for a specific element, e.g. aluminium to produce blue flowers, or is there likely to be a specific deficiency or toxicity, e.g. molybdenum deficiency in lettuce and boron toxicity in *Beloperone*?

Not only is it necessary to maintain a supply of each of the plant nutrients, but a balance must also be maintained between the concentrations of the different elements. The effects of individual elements are often discussed separately but large and important *interactions* occur between nutrients. High levels of some nutrients may accentuate deficiencies of other nutrients. For example, the application of nitrogen and potassium may cause an increased rate of growth, resulting in phosphorus deficiency, or the addition of large amounts of potassium to the compost may reduce the plants' uptake of magnesium, and although a reasonable quantity of magnesium is present in the compost, magnesium deficiency can still occur. This is known as a potassium-induced magnesium deficiency and is one example of antagonism between elements. These and other nutrient interactions are readily seen from factorial design experiments which allow both the main effects and the interactions to be observed. Experiments of this type yield much more information than simple trials, and their use has been advocated by R. A. Fisher (1926) who said, 'No aphorism is more frequently repeated in connection with field trials than that we must ask nature a few questions, or, ideally, one question at a time. The writer is convinced that this view is wholly mistaken. Nature, he suggests, will best respond to a logically and carefully thought out questionnaire, indeed if we ask her a single question, she will often refuse to answer until some other topic has been discussed.'

The functions in plant nutrition of the remaining five macro-elements and the form in which they can most conveniently be applied will now be examined, together with some of the most important of the interactions between the elements and the environment. The modern practice of referring to nutrients by element rather than oxide, e.g. K instead of K_2O, has been adopted in this book. The traditional oxide basis is often illogical, e.g. when stating the K_2O content of KCl, the oxide basis has only been retained when referring to other published work. Factors for converting from the element to the oxide basis are included in Appendix 6.

6.1 PHOSPHORUS

Chemical analysis of the tissue of a typical pot-grown plant will show that, by comparison with nitrogen and potassium, the amount of phosphorus absorbed is relatively low, often only one tenth of the amount of either of the other elements. This does not mean that less attention is required in the phosphorus nutrition of pot plants; indeed, maintaining an adequate and continuing supply of phosphorus in loamless composts is one of the most important aspects of macro-element nutrition.

Phosphorus is present in relatively large quantities in the protoplasm and nucleus of the cell. It is a component of the nucleic acids and sugars and is essential for many of the energy transfer processes, such as photosynthesis and the breakdown of carbohydrates, which occur in plants. Deficiency symptoms can include a dark green colour of the upper leaves, and the lower leaves turning a pale green with large yellow patches developing at a later stage. Phosphorus is one of the mobile elements within the plant, and, when there is a deficiency in the compost, the plant is able to withdraw some of the phosphorus from the lower leaves and transfer it to the young actively growing tissue. This is the reason why deficiency symptoms are first seen in the older leaves. Young tomato seedlings, deficient in phosphorus, will often show a reddish-purple colour on the under-sides of the cotyledons. This pigmentation can be caused by low air temperatures, but it is also known that low soil temperatures have a greater effect on reducing the absorption of phosphorus by the plant than on the absorption of potassium and other cations. High levels of phosphorus in composts are traditionally considered necessary for good root growth and the rapid establishment of young seedlings.

In mineral soils, the concentration of phosphorus in the soil solution is low by comparison with other macro-elements. It has been estimated that in order to maintain adequate growth rates the phosphorus in the soil solution has to be renewed several times each day to replace that taken up by the plants. Unlike mineral soils, which contain apatite and other minerals that provide a natural supply of phosphorus, the materials commonly used to make loamless composts do not have a natural supply of phosphorus, and neither do they have the ability to fix or retain phosphorus to the same extent as mineral soils. Materials used to make loamless composts contain relatively little aluminium and iron, and their capacity to fix phosphorus in the insoluble aluminium and iron forms can, from the pot-plant grower's point of view, be regarded as insignificant. More attention must therefore be given to maintaining adequate levels of phosphorus in these composts, than is the case with mineral soil composts. The high degree of water solubility of phosphorus in peat-sand composts, in contrast to that of a John Innes compost, was discussed in Chapter 4.

The chemical form in which phosphorus is present in composts is largely determined by the pH. Olsen (1953) concluded that, at pH 5, practically all the inorganic orthophosphorus present in the soil solution was in the form of $H_2PO_4^-$, and, as the pH was raised to 7·2, the phosphorus was equally divided between the $H_2PO_4^-$ and HPO_4^{--} forms. High pH values reduce the availability of phosphorus, and plants grown at a high pH contain much less phosphorus than those grown at a lower pH. The total amount of mineral phosphorus present in peats is low, being about 0·01 to 0·05% of the oven-dry weight; the amount of organic phosphorus which is mineralised and made available during the growth period of a typical pot-plant is also small. For all practical purposes, it must therefore be assumed that all of the plant's requirement for phosphorus must be met by the application of phosphatic fertilisers.

110   *Other Macro-Elements*

Although several phosphatic fertilisers are available, in practice only a few of them are used in preparing loamless composts and it is convenient to consider them under two groups: (a) those in which the phosphorus is water-soluble and (b) fertilisers having insoluble forms of phosphorus. It should be noted, however, that not all of the so-called water-soluble fertilisers are suitable for making liquid fertilisers. Whilst the phosphorus may be in a water-soluble form, other non-soluble compounds may be present, e.g. the calcium sulphate in superphosphate.

WATER-SOLUBLE PHOSPHATIC FERTILISERS
Superphosphate. This is the most widely known and used phosphatic fertiliser. It is produced by adding sulphuric acid to finely ground rock phosphate.

Fig. 6.1   The interaction of rates of superphosphate and the nitrogen source. *Top (left to right):* decreasing rates of superphosphate with urea as the nitrogen source. *Bottom (left to right):* decreasing rates of superphosphate with calcium nitrate as the nitrogen source.

This gives a water-soluble, monobasic calcium phosphate mixed with twice its weight of calcium sulphate. It normally has 8% water-soluble phosphorus with a little insoluble rock phosphate (1–2%) and some insoluble iron and aluminium phosphates; some micro-element impurities such as boron are also present. The fertiliser is available in powder or granular forms.

When the nitrogen in the base fertiliser is supplied in organic form, high rates of superphosphate often show marked beneficial effects on plant growth which cannot be accounted for simply on the grounds of increased levels of available phosphorus and an improved N:P ratio; this situation is demonstrated in Figure 6.1. When nitrogen was supplied in the nitrate form, plant growth responded little to varying the amount of superphosphate; when the nitrogen was supplied as urea, however, there was a progressive increase in growth as the amount of superphosphate was increased. Whilst superphosphate is regarded as having a neutral effect in agricultural soils, when used in a peat-sand compost, it will reduce the pH of the compost (Table 6.1). Bunt and Adams (1966, *b*) investigated the beneficial effect of

Table 6.1. *Effect of forms and rates of superphosphate on the compost[1] pH, determinations made in M/100 $CaCl_2$. (Figures in parentheses are for determinations made in water.)*

Fertiliser	Rate, kg/cu.m			
	Nil	0.75	1.50	3.00[2]
Single superphosphate (8%P)	5.22 (5.71)	4.87 (5.28)	4.57 (4.79)	4.39 (4.64)
Triple superphosphate (21%P) at equivalent rates of phosphorus	5.22 (5.71)	4.80 (5.35)	4.63 (5.20)	4.46 (4.87)

[1] Compost made from 75% Irish peat and 25% fine sand. Other fertilisers were used at the normal rate for the GCRI compost (p.156).

[2] These rates are approximately equivalent to 1, 2 and 4 oz./bushel or 1 lb. 5 oz., 2 lb. 10 oz., and 5 lb. 4 oz./cu. yd

using high rates of superphosphate, when the nitrogen source was hoof and horn, by following changes in the forms of nitrogen and the compost pH over a period of 8 weeks. They concluded that the probable reasons for the beneficial effect of the superphosphate on plant growth were the reduction in the compost pH, which also slowed down the rate of organic nitrogen mineralisation, a 50% reduction in the amount of free ammonia in the compost, and also a reduction in the amount of nitrite-nitrogen produced.

Triple superphosphate. As the name suggests, this fertiliser contains about three times as much phosphorus as single superphosphate. It is made by using phosphoric acid in place of sulphuric acid to dissolve the rock phosphate. The fertiliser has 21% water soluble phosphorus and, unlike single superphosphate, it does not contain any significant amount of calcium sulphate. It is available in both powder and granular forms.

Monoammonium phosphate ($NH_4H_2PO_4$). The agricultural grade of this fertiliser, which is known as Ammo-Phos A in the USA, contains 21% P and 11% N. It is made by combining ammonia with a low grade of phosphoric acid obtained from the wet process in which rock phosphate is treated with sulphuric acid. In horticulture, monoammonium phosphate is used chiefly as a phosphorus source in liquid feeds, and, to minimise the formation of precipitates, it is essential that only the high grade fertiliser (p.191) made from thermal process phosphoric acid is used. In the thermal process of phosphoric acid manufacture, elemental phosphorus is first produced from rock phosphate in an electric-arc furnace and then oxidised to P_2O_5 with air. This method of production gives a material with very low amounts of such impurities as iron and aluminium which can cause precipitation problems. Monoammonium phosphate made from this grade of phosphoric acid has an analysis of 25% P and 12% N.

Diammonium phosphate (($NH_4)_2HPO_4$). This is also known as Diammonphos; it contains 23% P and 21% N.

Both of the ammonium phosphates are being increasingly used in high-analysis base fertilisers for agricultural purposes where their higher nutrient content and lower handling costs make them competitive with the lower analysis phosphorus fertilisers. The analysis of these two fertilisers varies slightly with the method used in their manufacture.

Phosphoric acid (H_3PO_4). A caustic liquid containing 24% P. It is sometimes used by fertiliser manufacturers as a phosphorus source in liquid feeds and is also applied in the irrigation water for some field crops in western parts of the USA. Some care in its handling is required, if it is used by growers in the preparation of liquid feeds on the nursery. (See monoammonium phosphate, regarding suitable grades of material.)

Superphosphoric acid. This is the most concentrated form of phosphorus available, containing 33% P. It is primarily used in the formulation of commercial liquid fertilisers; approximately half the phosphorus is in the normal orthophosphoric (H_3PO_4) form, and the remainder is largely present as pyrophosphoric acid ($H_4P_2O_7$). This latter form can sequester or chelate metals such as iron, aluminium, copper, zinc and manganese which would otherwise be precipitated as phosphates and so reduce the solubility of the fertiliser. This chelating action makes it a useful source of phosphorus for liquid fertilisers.

Polyphosphate. When superphosphoric acid is neutralised with ammonia, it gives a highly concentrated water-soluble fertiliser containing 15% N and 27% P; 50% of the phosphorus is in the ortho form and about 50% in the polyphosphate form, mostly as triammonium pyrophosphate [$(NH_4)_3H_4P_2O_7$]. The polyphosphates are hydrolised in the soil to orthophosphate, before being absorbed by plants.

SLOW RELEASE OR INSOLUBLE FORMS OF PHOSPHORUS

Phosphatic fertilisers are also available in insoluble forms; rock phosphate and basic slag are traditional forms of insoluble phosphate fertilisers and are widely used in agriculture. Other materials such as the metal ammonium phosphates and the metaphosphates have recently been used in loamless composts as a means of supplying phosphorus over a prolonged period.

Basic slag. A by-product from the manufacture of steel from iron ores rich in phosphates, the slag is finely ground and contains 3–9% water-insoluble phosphorus which is largely present as tetracalcium phosphate ($Ca_4P_2O_9$), with some silicophosphate. This wide variability in phosphorus content, together with its strong basicity, equivalent to about 60% by weight of calcium carbonate, does not generally make it a suitable phosphorus-supplying fertiliser for horticultural purposes; its only appeal lies in supplying phosphorus in an insoluble form.

Rock phosphates. Finely ground rock phosphates, containing 11–13% phosphorus, are another form of slowly available phosphorus. Traditionally used on grasslands, their effectiveness and composition vary with the source of the material. A fine particle size, a high organic content of the soil and a low pH are factors generally believed to be important for the utilisation of this form of phosphorus.

Other traditional slow-release phosphatic fertilisers, such as bone meal and steamed bone flour, have variable amounts of phosphorus and nitrogen; they are relatively expensive and do not offer any particular advantages over other phosphatic fertilisers.

Magnesium ammonium phosphates ($MgNH_4PO_4 \cdot H_2O$). Fertilisers in this group are of variable composition. A typical analysis would be 8% N, 20% P, 13% Mg; potassium can also be included to give a combined magnesium-ammonium and potassium-ammonium phosphate. Some aspects of this group of fertilisers have already been discussed in Chapter 5, with reference to their nitrogen supplying characteristics. The phosphorus rate of availability from this group of fertilisers is such that, for those pot plants grown over a three month period, e.g. chrysanthemum and poinsettia, sufficient fertiliser can be added when mixing the compost to meet the total phosphorus requirements of the crop without the need to include phosphorus in the liquid feeds. Under certain conditions, the inclusion of soluble phosphorus in the liquid feeds can lead to the formation of precipitates, resulting in the blockage of irrigation lines (see Chapters 9 and 10). The rate of phosphorus availability from this group of fertilisers can be regulated by the size of the granules: for use in potting composts 80% of the granules should be between 1·5 and 3·0 mm in diameter. The main potential use of these materials appears to be as slow-release sources of phosphorus, rather than nitrogen sources as originally recommended.

Calcium metaphosphate $(Ca(PO_3)_2)$. Calcium metaphosphate is a glass-like material produced by burning elemental phosphorus in a combustion chamber into which finely ground rock phosphate is injected; it is tapped from the furnace as molten glass and then cooled and crushed. This gives a material having 27–28% phosphorus, about three-quarters of which is derived from the elemental phosphorus and the remainder from the rock phosphate; it also has 25% of calcium oxide (Nelson, 1965). The material is only slightly soluble in water but is sufficiently soluble in the weak acids present in the soil to give a high rate of availability to plants.

Potassium metaphosphate (KPO_3). Whilst this material contains slightly less phosphorus (25% P) than potassium (32% K), its prime interest lies in increasing the range of slow-release phosphorus sources. It has a low water-solubility but hydrolises in the compost to give the more soluble orthophosphates, the rate of hydrolysis being affected by the moisture content, the temperature and acidity of the compost.

To date, these two metaphosphates have received wider attention for agricultural crops than they have for ornamentals, and there is insufficient data upon which to base recommendations for potting composts. Their resistance to leaching and their rates of phosphorus release suggest that they have a large potential use in pot plant composts.

6.2 POTASSIUM

The amount of potassium present in the tissue of pot plants is often of the same magnitude as that of nitrogen, an average value being 3–4% of the dry weight. When plants are grown with high levels of potassium in the compost, however, the amount of potassium present in the plant tissue can be very high: this condition is often referred to as 'luxury consumption'.

Potassium originates from such minerals as the micas and felspars, and an average value for potassium in a mineral soil would be about 2%. The amount of potassium present in organic materials is usually much lower than that in mineral soils. An average value for peat would be about 0·04% of the oven-dry weight, and when an allowance has been made for the very much lower bulk density (about 0·1–0·2 g cc^{-3}) for peat, by comparison with 1·2 g cc^{-3} for a mineral soil, it will be readily seen that loamless composts based upon peat will have a very low natural content of potassium. Vermiculite is the only material used in making loamless composts which contains significant amounts of potassium. In composts made with materials having a base exchange capacity, the potassium will be present in both the water soluble and exchangeable forms. As the potassium present in the soil solution is either absorbed by the plant or lost by leaching, some of the exchangeable form is released to take its place. Similarly, when water soluble potassium is applied in the liquid feed, some of it will be taken into the exchangeable form.

Potassium is necessary for the formation of enzymes that control carbohydrate and nitrogen metabolism; it also acts as an osmotic regulator in the water relations of plants. Horticulturists normally associate it with balancing the soft growth resulting from too much nitrogen; such plants are generally considered as being more susceptible to fungal diseases. It is also essential for the production of high quality tomato fruit, where large quantities of potassium are supplied in relation to the amount of nitrogen. Often the liquid fertilisers applied to tomatoes have a ratio of 1N to 2K, whereas for poinsettia culture the reverse ratio of 2N to 1K would be more usual. Potassium is also believed to play a role in controlling the water relations within the plant; there is some evidence that the plant can partially substitute sodium for potassium.

Deficiency symptoms in plants often start with mottling of the foliage and pale leaf margins. As the deficiency increases, necrotic spots develop, the leaf edges scorch and turn brown, and the lower leaves may absciss.

POTASSIUM FERTILISERS

Potassium does not have such complex reactions within the compost as either nitrogen or phosphorus. Only relatively few chemicals are suitable for use as potassium-supplying fertilisers and, with few exceptions, they do not present any problems.

Potassium sulphate (K_2SO_4). This fertiliser contains 44·8% K; it is a dry powder which does not readily absorb water vapour from the atmosphere and is easy to handle. Whilst the commercial grade of fertiliser supplies potassium in a water soluble form, the impurities present prevent it being used to prepare the highly concentrated liquid fertiliser stock solutions which are diluted before being applied to the plants. The commercial grade fertiliser can be used only as a base fertiliser for composts; a more refined grade containing less impurities and having a higher solubility must be used if it is desired to use potassium sulphate in the liquid feed.

Potassium chloride (KCl). Also known as muriate of potash, this material contains 51% K and has a high degree of water solubility. It contains 47% chloride, and because of the risk of chloride toxicity is not generally used in Britain for glasshouse crops; it has, however, been the most widely used potassic fertiliser in the USA. Kainit or Potash Salts (12·5–16% K) also contain chlorides and for this reason should be avoided.

Potassium carbonate (K_2CO_3). This material contains 55% K; it is basic in its reaction and can be used where it is desired to raise the pH of the compost. It is not in common use in Britain.

Potassium nitrate (KNO_3). In addition to supplying 38·6% K, this fertiliser also contains 13% of nitrate-nitrogen. It has a high degree of water-solubility which makes it the obvious choice of a potassic fertiliser for preparing liquid feeds, providing that the amount and form of nitrogen is acceptable. It can also be used in the base fertiliser but is slightly hygroscopic and for this reason must be stored under dry conditions.

Potassium frit. The term 'frit' or 'fritted' refers to the process of fusing a potassium salt with molten sodium silicate. This is rapidly cooled by pouring it either onto a cold surface or into running water where it solidifies and shatters into small pieces which are then finely ground. This gives a material that has 29% K in a form which is not immediately water soluble, but which dissolves slowly in the compost and releases potassium over a long period. This form of slow-release action is not usually considered necessary in the case of potassium, where the initial concentration in the compost can be readily controlled by regulating the strength of the base fertiliser. The potassium level can then be easily maintained by adjusting the strength of the liquid fertiliser. The fritted type of fertiliser has a much greater potential for supplying micro- or trace elements (Chapter 7).

6.3 CALCIUM

This element is required by the plant in regions of active cell division for the formation of cell walls in which it occurs as calcium pectate. It also plays a role in the transport of carbohydrates and amino acids within the plant, and in the development of new roots.

A deficiency of calcium shows as a stunting of plant growth. Leaves are restricted in their development and can show a paleness at the margins, which may be followed by an inward curling. These leaf symptoms, which are most noticeable in broad-leafed plants, may be followed by necrotic spotting. In cases of acute deficiency, the tips of the plants will die and lateral buds remain dormant. Roots are particularly sensitive to calcium deficiency; young roots and root hairs are killed and the older roots turn brown. Wind (1968) found that very little root growth occurred in peat when the pH was below 3·5 (measured in 0·1 N KCl). Gammon (1957) studied the growth of roots in soils adjusted to different pH values by the use of carbonates of calcium, sodium or potassium, and concluded that the lack of calcium was more detrimental to root growth than the pH itself. Calcium is also required for fruit and seed development, a good example of calcium deficiency being blossom-end rot in tomatoes. This is usually seen at the distal or blossom end of the fruit as a water soaked region, which develops into blackened, hard dry areas; symptoms may also be present within the fruit but not visible from the outside. This condition can be induced by high salt stress, dryness and high levels of soluble potassium, magnesium and ammonium, all of which reduce the uptake of calcium by the plant.

Apart from their effects upon the soil salinity and certain antagonistic effects towards one another, most elements can be regarded as having only one prime function, i.e. their direct role in plant nutrition. Calcium, however, also has an important indirect role. It is the element normally found in the greatest concentration in the compost and has a greater effect in controlling the reaction or pH of the compost than any other element. In addition to its removal in relatively large quantities by the plant, usually of the same order of magnitude as for nitrogen and potassium, it is also lost in large amounts by leaching. The fertiliser, ammonium sulphate, is generally regarded as causing a loss of calcium carbonate equal to its own weight, and this is followed by a drop in the pH.

Calcium is therefore important in plant nutrition because of:

(1) its function within the plant;
(2) its effect on the pH and thereby on the availability of other macro- and micro-elements;
(3) its effect on disease control, e.g. liming to a high pH helps to control clubroot in brassica and fusarium in tomato, whilst a pH of not greater than 5·0 is considered helpful in controlling *Thielaviopsis* root rot in poinsettia.

The quantity of calcium required to change the pH is related to the buffer capacity and cation exchange capacity of the compost (see p. 73).

CALCIUM-SUPPLYING MATERIALS

The most common forms of calcium or liming materials are calcium oxide, calcium hydroxide and calcium carbonate.

Calcium oxide (CaO). Commonly known as burnt lime, quicklime or unslaked lime, it is obtained by heating calcium carbonate and driving off the carbon dioxide. This gives a highly concentrated form of lime which can be caustic to the roots of young seedlings; it can also be disagreeable to handle. It is often used in agriculture but is not recommended for use in potting composts.

Calcium hydroxide ($Ca(OH)_2$). Also known as slaked lime, quenched lime or hydrated lime, this form has a solubility of 1 700 ppm in water and can also have a caustic action on young seedlings. It used as a liming material in composts, an interval of several days should elapse between mixing and using the compost. Its appeal lies in achieving a more rapid increase in the pH of the compost than is obtained with chalk; also, being in a more concentrated form than chalk, less material is required. Both calcium oxide and calcium hydroxide eventually revert to calcium carbonate in the compost.

Calcium carbonate ($CaCO_3$). This material is also known in horticultural circles as chalk, carbonate of lime and ground limestone. It is an inherently safe form of lime with no caustic action, and is the form recommended for use in potting composts. Its solubility in pure water is only 10 ppm but this increases to 600 ppm when the water contains carbon dioxide in solution. Its effectiveness or speed of action in the compost is largely determined by its particle size; the finer the particles, the more rapid its effect. For use in potting composts not less than 90% by weight should pass a 100 mesh sieve.

The efficiency of these materials for liming purposes can be compared from their molecular weights, and this is often referred to as their neutralising

Table 6.2. *The relative neutralising value of various liming materials.*

Liming Material	Molecular weight	Neutralising value	
		$CaCO_3 = 100\%$	$CaO = 100\%$
Calcium oxide (CaO)	56	178	100
Calcium hydroxide ($Ca(OH)_2$)	74	135	75
Calcium carbonate ($CaCO_3$)	100	100	56
Magnesium oxide (MgO)	40	250	140
Magnesium hydroxide ($Mg(OH)_2$)	58	172	96
Magnesium carbonate ($MgCO_3$)	84	119	97

value; unfortunately, some countries adopt calcium carbonate ($CaCO_3$) as the reference value, whilst others use calcium oxide (CaO). The comparative neutralising values of various liming materials are shown in Table 6.2. Other materials, such as wood-ash and basic slag, also have some neutralising value, but they are not standardised materials and are not to be recommended.

Calcium Sulphate ($CaSO_4$). Commonly known as gypsum, this can be used whenever it is desired to add calcium to the compost without appreciably affecting the pH value; only under very acid conditions will calcium sulphate raise the pH.

6.4 MAGNESIUM

Magnesium is one of the elements required for the formation of chlorophyll and a deficiency will soon lead to a reduction in the rate of photosynthesis. It also has an effect on the transport of phosphorus within the plant.

Commonly observed deficiency symptoms are a yellow-green mottle, usually appearing first on the older leaves, the veins remaining green. In some plants a deep red pigment develops in the centre of the leaves and this may be followed by necrotic spots; this symptom is frequently seen in pot-grown tomatoes. Magnesium deficiency may be due to low levels of exchangeable magnesium in the compost but often it is the result of very high levels of potassium rather than a true deficiency of magnesium. Branson *et al.* (1968) investigated this problem in chrysanthemums and found that, at an early stage in the life of the crop, the leaves had 0·1% Mg and 4·8% K (dry weight basis); as the crop developed the magnesium in the leaves fell to 0·04% and deficiency symptoms appeared. Adding magnesium sulphate to the soil was not found to be very effective in reducing the deficiency unless the feeding of potassium was discontinued.

Experiments at the Glasshouse Crops Research Institute have shown that magnesium deficiency symptoms are unlikely to occur when the K:Mg ratio in the compost is 3:1 or less. Hooper (1973) has also concluded from the examination of large numbers of composts for advisory purposes that plants are free from deficiency symptoms when the ratio of extractable K:Mg is below 3:1. At a ratio of 4:1 and above, the risk of deficiency symptoms was found to be much greater.

Magnesium deficiency in crops such as tomato and chrysanthemum can be controlled also by applying a foliar spray; usually a 2% solution of Epsom salts is used. There is some evidence that the rate of magnesium absorption by the leaves is faster with magnesium nitrate and magnesium chloride than with magnesium sulphate. The leaves of some plants can be scorched if solutions stronger than 2% are used.

MAGNESIUM SUPPLYING FERTILISERS

Although several magnesium supplying materials are available, normally only two forms are in common use in potting composts.

Magnesium sulphate. This is available as Epsom salts ($MgSO_4 \cdot 7H_2O$), with 10% Mg, and as Kieserite ($MgSO_4 \cdot H_2O$) which has less water of crystallisation and therefore a higher magnesium content (16%). Kieserite is also cheaper than Epsom salts. Magnesium salts can sometimes cause toxicity to plants and it is usually safer to use magnesium limestone in composts.

Magnesium limestone. This material is a calcium–magnesium carbonate containing 11–13% Mg; a low grade material having only 3% Mg is also available. The terms Dolomite limestone and Dolomitic limestone are often incorrectly used synonymously with magnesium limestone. In correct usage, Dolomite refers to a compound containing equal amounts of calcium and magnesium carbonate with 12% Mg. When the limestone contains very little magnesium it is termed *calcic,* as the magnesium content of the material rises it is called *Dolomitic,* and when the material is composed almost entirely of calcium-magnesium carbonate it is then called *Dolomite,* Thus, in practice, *Dolomitic* limestone has a lower magnesium content than *Dolomite* limestone.

In addition to its function of supplying magnesium, this material is also used to neutralise soil acidity. By laboratory methods, it usually shows approximately the same neutralising value as calcium carbonate, but, under practical conditions, it is somewhat slower than calcium carbonate in raising the compost pH, as magnesium carbonate is less easily dissolved by mineral acids than calcium carbonate. In some situations, however, magnesium carbonate is preferable as a liming material to calcium carbonate. Tod (1956) grew *Rhododendron davidsonianum* in a soil, to which large amounts of magnesium carbonate was added. This raised the pH to levels which would have been considered toxic to rhododendrons if calcium carbonate had been used. No serious ill effects, however, resulted from its use and it was concluded that alkalinity *per se* was not the cause of the harmful effects produced by alkaline soils on rhododendrons.

Magnesium limestone is also available in the burnt form; the calcium and magnesium are then present as oxides. Magnesium oxide is less soluble than calcium oxide, however, and remains in a caustic condition for a longer period; for this reason burnt magnesium limestone is not recommended for use in potting composts.

Experiments made at the Glasshouse Crops Research Institute have shown that large interactions occur between the forms of the liming materials and the sources of nitrogen (Fig. 6.2). Using urea as the nitrogen source, Dolomite limestone was superior to calcium carbonate but was inferior to calcium

Magnesium 121

Fig. 6.2 The interaction of nitrogen source and liming material. *Top:* urea as the nitrogen source. *Bottom:* calcium nitrate as the nitrogen source. *Left:* calcium carbonate. *Centre:* Dolomite limestone. *Right:* calcium sulphate.

sulphate. With calcium nitrate as the nitrogen source, plant growth was similar with all three liming materials. The toxic effect of the urea was related to the pH of the compost; the higher pH obtained with the calcium carbonate caused the production of free ammonia and nitrite nitrogen. Toxicities due to these forms of nitrogen have been discussed in Chapter 5. Dolomite limestone, used at the same rate as calcium carbonate, gave a slightly lower compost pH and the growth depression was correspondingly reduced; calcium sulphate had the least effect upon the compost pH and gave the best growth. These results illustrate the importance of achieving the correct balance in plant nutrition and also the need to exercise caution in changing the formulation of a compost. Simply substituting one form of nitrogen or liming material for another can often have significant effects on plant growth, unless the possible interactions of the fertilisers are first considered.

Peats differ from mineral soils in having a much greater percentage of magnesium present in an exchangeable form, and, providing that some magnesium has been included in the base fertiliser, deficiency symptoms are not liable to occur, unless the potassium level is allowed to rise excessively. One plant which is very susceptible to magnesium deficiency induced by high levels of potassium is the winter cherry (*Solanum capricastrum*). Owen (1948) investigated this problem with regard to potting composts based on a mineral soil. The most effective treatment was the addition of Epsom salts (magnesium sulphate) in the base fertiliser at rates equal to the amount of sulphate of potash used. If the potassium had been supplied by a compound or complete N P K fertiliser, then the amount of magnesium sulphate should be equal to twice the amount of K_2O in the compound fertiliser.

6.5 SULPHUR

This element is an important part of the amino acids or nitrogenous compounds from which proteins are formed. Plants are unable to utilise sulphur in the elemental form; it must be absorbed by the roots as sulphates. Sulphur deficiency symptoms are seldom seen in pot-grown plants which are normally adequately supplied with sulphates from fertilisers such as superphosphate, ammonium sulphate, etc., and also as an impurity in the irrigation water; an excess of sulphate is usually a greater problem than a deficiency. Fortunately, plants are more tolerant of an excess of sulphate than they are of chloride. Eaton (1942) found that chlorides were approximately twice as toxic as sulphates to the tomato, bean and lemon. Saturated solutions of calcium sulphate have been applied to primula, carnation, chrysanthemum, fuchsia and poinsettia without any harmful effect; azaleas and Rex begonias, however, have shown unfavourable reactions (Pearson, 1949). High concentrations of sulphates in the compost present problems when salinity determinations are made using wide-ratio suspensions of compost to water, rather than saturated paste extracts. The precautions necessary in making salinity determinations under these conditions have been discussed on pages 78-81.

By comparison with mineral soils, peats often have a relatively high sulphur content which oxidises to sulphates when peats are drained and cultivated, and deposits of calcium sulphate can often be seen in the drainage ditches of peat bogs.

6.6 MINERAL SOIL AND PEAT COMPARISON

The quantity of each of the macro-elements present in the various types of peats and mineral soils in their natural states can be very variable, being dependent upon the basic types of vegetation and minerals from which the peats and the soils have been formed. Simplified comparisons of the mineral values of these two media must, therefore, be treated with some caution and the values given in Table 6.3 should be regarded only as showing the large differences that can exist between the two types of media. It must also be noted that the values in the table are on a weight basis, and the bulk density of peat is only 1/10 or less of that of a mineral soil. When the nutrient values are transformed to a volume basis, they will be reduced correspondingly. Apart

Table 6.3. *Comparative mineral values of an average peat and mineral soil, % by weight. N.B.–The bulk density of peat is only approximately one tenth that of mineral soil.*

Nutrient	Peat	Mineral soil
Organic matter	95·0	3·0
C/N ratio	30:1	10:1
Nitrogen	2·5	0·15
Phosphorus	0·03	0·05
Potassium	0·04	1·80
Calcium	0·20	0·35
Magnesium	0·15	0·30
Sulphur	0·15	0·04

from the sulphur and the organic matter, peats will be seen to have a much lower content of plant nutrients than mineral soils. We have already noted, however, that when fertilisers are added to the two media at equal rates per *unit volume,* the levels of available plant nutrients in the peat-based composts can be expected to be higher than those of mineral soil composts. This is because of the lower amount of 'fixation' that occurs.

6.7 NUTRIENT AND ENVIRONMENT INTERACTIONS

In studies at the Glasshouse Crops Research Institute on the nutrition of plants grown in peat-based composts, several interactions have been observed between certain macro-elements; interactions have also occurred between the nutrition and the environment or the season.

An example of an interaction between macro-nutrients is shown in Figure 6.3. Antirrhinum was grown in peat-sand composts having either calcium nitrate or hoof and horn as the nitrogen source; with each fertiliser there were two rates of calcium carbonate application. At the high rate (pH 6·0 in $M/100$ $CaCl_2$), only calcium nitrate was an acceptable source of nitrogen; the hoof and horn treatment showed marked phytotoxicity and growth depression. At the lower rate (pH 4·9), there was no significant difference in growth between plants in the two sources of nitrogen.

124 *Other Macro-Elements*

An example of the type of interaction which can occur between the nutrition and the environment, where tomatoes were grown in two factorial design experiments, is seen in Figure 6.4. The four treatments had received calcium carbonate and nitrogen at either low or high rates of application, the nitrogen being 70% in the form of hoof and horn, and 30% as ammonium nitrate. Other nutrients were given at constant rates for all the treatments. In winter, there

Fig. 6.3 The interaction of nitrogen source and rate of calcium carbonate. *Left to right:* calcium nitrate, high rate of calcium carbonate; hoof and horn, high rate of calcium carbonate; calcium nitrate, low rate of calcium carbonate; hoof and horn, low rate of calcium carbonate.

were large differences in growth between the four treatments. Plants having the high rate of nitrogen and the high rate of calcium carbonate application made the least growth; often this treatment resulted in the eventual death of the plants. With plants at a comparable stage of development in summer, however, there was very little difference in growth between the four treatments, and eventually the high nitrogen treatments produced the most growth. The beneficial effect of the high rate of superphosphate, as shown in Figure 6.1, is also much greater in winter than it is in summer. The requirement of phosphorus by the plants, however, is actually lower in winter than it is in summer, because of their lower growth rate under the low light conditions of winter. Some of the reasons for this beneficial effect of superphosphate have been discussed on page 110. Because of the higher compost temperatures, the rates of mineralisation of hoof and horn and some other forms of slow release fertilisers are approximately twice as high in summer as in winter. The rates of plant growth, however, can be ten or more times greater in summer than winter; consequently there is less risk of nitrogen toxicity. True mineral deficiencies are more quickly observed in summer than in winter.

Nutrient and Environment Interactions 125

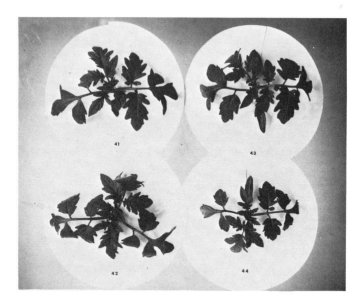

Fig. 6.4 The interaction of the rate of organic nitrogen, the rate of calcium carbonate and the season. *Top illustration:* tomatoes grown in four composts from a factorial experiment in winter. *Top left:* low nitrogen, low calcium carbonate. *Top right:* high nitrogen, low calcium carbonate. *Bottom left:* low nitrogen, high calcium carbonate. *Bottom right:* high nitrogen, high calcium carbonate.
Lower illustrations: the same treatments, grown in summer.

6.8 FERTILISER ANALYSIS AND SALT INDEX

A summary of the most commonly used fertilisers and their principal plant nutrient contents is given in Table 6.4. The nutrient content may vary slightly in different countries, depending upon the precise method of fertiliser manufacture and also the sources and purity of the basic materials used.

All soluble fertilisers contribute towards the salinity of the soil solution, some more than others. As already seen (Chapter 4), high salinity values are

Table 6.4. *Nutrient analysis of the fertilisers commonly used in preparing loamless composts.*

Fertiliser	Analysis				
	N	P	K	Ca	Mg
Ammonium nitrate	35·0				
Ammonium sulphate	20·5				
Calcium nitrate	15·5			20·0	
Sodium nitrate	16·0				
Urea	46·0				
Basic slag		3–9		32	
Calcium metaphosphate		27·5		17·8	
Diammonium phosphate	21	23			
Monoammonium phosphate	12	25			
Superphosphate (single)		8–10		17–22	
Superphosphate (triple)		21			
Ammonium polyphosphate	15	27			
Phosphoric acid		24			
Superphosphoric acid		33			
Potassium carbonate			55		
Potassium chloride			49		
Potassium metaphosphate		25	32		
Potassium nitrate	13·8		38		
Potassium sulphate			41		
Calcium carbonate				40	
Calcium hydroxide				54	
Calcium oxide				71	
Calcium sulphate				57	
Magnesium carbonate					11–13
Magnesium hydroxide					41
Magnesium oxide					60
Magnesium sulphate (Epsom salts)					10
Kieserite					16

detrimental to plant growth and the aim should be to choose a fertiliser which, for a given amount of plant nutrients, has the least effect on soil salinity. Rader *et al.* (1943) determined the effects of *equal weights of fertilisers* upon the osmotic pressure of the soil solution, these data are given in Table 6.5. To enable the more useful comparisons to be made of the effect of *equal amounts of nutrients*, which is the basis used in formulating potting composts, the data have been recalculated to give the salt index of the fertiliser relative to its N, P or K content. In both cases the results have been expressed relative to the effect of sodium nitrate. The actual increase in osmotic pressure from a given amount of fertiliser will depend upon the compost, principally the amount of water and colloidal material it contains. It will be seen that on the *equal*

Table 6.5. *Effect of fertilisers on the soil salinity.*

Fertiliser		Salt[1] Index	Total[2] Nutrients	Relative[3] Salinity
Sodium nitrate	16·5 N	100	16·5	100
Ammonium nitrate	35 N	104·7	35·0	49·4
Ammonium sulphate	21 N	69·0	21·0	53·7
Ammonia solution	82 N	47·1	82·0	9·4
Calcium nitrate	11·9 N	52·5	28·8	30·1
Urea	46 N	75·4	46	26·7
Diammonium phosphate	21 N, 23 P	34·2	44	12·7
Monoammonium phosphate	12 N, 27 P	29·9	39	12·7
Superphosphate (single)	7·8 P	7·8	7·8	16·5
Superphosphate (triple)	19·6 P	10·1	19·6	8·5
Potassium chloride	49·8 K	116·3	49·8	38·5
Potassium nitrate	13 N, 38 K	73·6	51	23·6
Potassium sulphate	45 K	46·1	45	17·0
Kanit	14·5 K	109·4	14·5	124·5
Calcium carbonate	40 Ca	4·7	40	1·9
Calcium sulphate	23 Ca	8·1	23	5·8
Magnesium oxide	60 Mg	1·7	60	0·5
Magnesium sulphate (Kieserite)	16 Mg	44·0	16	44·5
Dolomite	24Ca 12Mg	0·8	36	0·4

[1] The 'Salt Index' was calculated from the increase in osmotic pressure of *equal weights* of fertilisers.

[2] 'Total Nutrients' have been recalculated from the sum of the N, P, K, Ca and Mg as usually stated in the fertiliser analysis, e.g. monoammonium phosphate = 12 N + 27 P = 39, but superphosphate = 7·8 P. Superphosphate is not usually regarded as being a source of plant calcium.

[3] 'Relative Salinity' has been calculated from the increase in osmotic pressure *per unit of plant nutrient*.

weight basis, sodium nitrate and ammonium nitrate have approximately equal effects, with values of 100 and 104·7 respectively. When, however, comparison is made on an *equal nutrient basis*, which in this instance is nitrogen, then ammonium nitrate with a value of 49·4 has only half the salinity effect of sodium nitrate with a value of 100.

In addition to their direct effects on plant growth by way of the increase in the osmotic pressure or salinity of the compost, some fertilisers may be less desirable than others because of an associated effect or the risk of a specific toxicity, e.g. chlorides in the case of potassium chloride, and free ammonia and nitrites in the case of urea and anhydrous ammonia.

6.9 PLANT MINERAL LEVELS

Horticulturists have long been accustomed to having their composts and glasshouse border soils analysed at regular intervals, and the results used as the basis for subsequent nutrition and management. More recently, the value of knowing the levels of the macro-elements within the plant, usually of the leaf tissue, has been recognised. The advantages of this technique are:

(1) Positive identification of mineral deficiencies and toxicities can be made.
(2) Corrective treatment by way of liquid feeding or foliar sprays can be made often before the symptoms are fully developed and the plant is showing damage.
(3) Nutrient levels in the tissue are more stable than those in the compost. Whereas the level of nutrients in the compost can show wide fluctuations because of the frequency and strength at which liquid fertilisers are applied, the level of macro-elements in the leaves changes more slowly.

The limitations of this technique are:

(1) Normal or 'optimal' levels of minerals in the leaf tissue have not yet been established for the full range of ornamental crops.
(2) The amount of minerals present in the leaf is known to vary with the position of the leaf on the plant. Some elements, such as nitrogen, phosphorus, potassium and magnesium, are mobile, and when deficiencies occur in the compost, the plant is able to transfer these elements from the older leaves to the regions of active growth. Other elements, such as calcium, iron and boron, are immobile. Deficiencies of these elements are therefore first found in the regions of new growth.
(3) Differences in the 'optimal' mineral levels can vary also with the season, the age of the plant and the variety, as shown by the work of Nelson and Boodley (1966) with the carnation. For these reasons, there must be careful standardisation of the tissue chosen for analysis; usually the fifth and sixth leaves from the growing point are used.

Table 6.6. *Suggested levels of macro-elements in the leaves of certain ornamental plants (% by weight of dry tissue). Values, below which deficiency symptoms may start to appear, are given in parentheses.*

Crop	N	P	K	Ca	Mg
Carnation	3·0–5·0	0·25–0·45	2·5–4·0	1·0–2·0	0·2–0·5
	(<3·0)	(<0·05)	(<2·0)	(<0·6)	(<0·15)
Rose	3·2–4·5	0·20–0·30	1·8–3·0	1·0–1·5	0·25–0·35
	(<3·0)	(<0·20)	(<1·8)	(<1·0)	(<0·25)
Chrysanthemum	3·5–5·0	0·35–0·60	4·0–6·0	1·0–1·5	0·45–0·60
	(<3·0)	(<0·20)	(<2·0)	(<0·3)	(<0·2)
Poinsettia	4·5	0·9	2·8	0·75	0·57
Geranium	3·3–4·8	0·4–0·65	2·5–4·3	0·81–1·2	0·2–0·52
	(<2·4)	(<0·28)	(<0·62)	(<0·77)	(<0·14)
Hydrangea					
(Blue)	2·09	0·54	2·0	0·99	
(Pink)	2·36	1·98	2·04	1·43	
Azalea	2·0–2·3	0·3–0·5	0·8–1·0	0·2–0·3	0·17–0·33
	(<1·7)	(<0·16)	(<0·7)	(<0·16)	(<0·08)
Philodendron	2·0–3·5	0·2–0·35	3·0–4·5	0·35–1·0	0·25–0·50
Sansevieria	1·5–3·0	0·15–0·30	2·0–3·5	0·75–1·25	0·25–0·50

The levels of macro-nutrients which have been found to be satisfactory or optimal for a range of plants are given in Table 6.6. They are largely based on the work of Kiplinger and Tayama (1970), Joiner and Waters (1973), and the review of tissue analysis of some ornamental plants by Criley and Carlson (1970).

Chapter 7

Micro-elements

Elements which are required by plants in only very small amounts are known as *micro-* or *trace* elements. Although they are present in plants at only 1/1000 to 1/10000 of the amount of such macro-elements as nitrogen and potassium, they are nevertheless equally essential for normal plant growth. Deficiency of a micro-element such as molybdenum can be just as important as a macro-element deficiency. The elements which are generally regarded as being essential micro-elements are boron, chlorine, copper, iron, manganese, molybdenum and zinc.

Composts based on mineral soils do not normally require any micro-element additions, whereas those based on peat are often deficient in boron, copper, iron and molybdenum. The extent to which micro-element deficiencies occur in peat-based composts depends upon the species of plant being grown and also the type and source of peat, e.g. molybdenum deficiency occurs more frequently with lettuce and cauliflower than it does with chrysanthemum, whilst boron deficiency is seen more frequently in plants grown in sphagnum than in sedge peats.

Plants are generally much less tolerant of high levels of micro-elements than they are of macro-elements. For example, the amount of potassium in the compost can vary by several hundred milligrams per litre before plants show marked reactions, whereas with the micro-element boron, deficiency can occur below 0·5 mg/l. and toxicity at concentrations above 3 mg/l.; copper, manganese and zinc can also be toxic at relatively low concentrations. In composts made from materials having a cation exchange capacity, the cations copper, zinc and manganese behave in a similar way to other macro-nutrient cations such as calcium and potassium, whilst the anions boron and molybdenum behave similarly to phosphorus. With the exception of molybdenum, the availability of the micro-elements to plants decreases as the compost pH is increased; the reverse effect is obtained with molybdenum. The effect of the pH on the availabiltiy of micro-elements in organic soils has been investigated by Lucas and Davis (1962) and their results have been summarised in Figure 4.6 (p. 72). They concluded that, for organic soils, the pH should be about 5·0 to 5·5; this is 1 to 1·5 pH units lower than the optimal value for mineral soils.

The choice of a solvent to estimate the quantity of micro-elements, which are available to plants from the compost, is more critical than it is for macro-element estimations. Mitchell (1971) recommends that the following solvents give the best results in relation to plant uptake:

Element	Solvent
Boron	Hot water
Molybdenum	Neutral normal ammonium acetate
Copper	0·05 Molar EDTA
Zinc	0·5 Normal acetic acid
Manganese	Neutral normal ammonium acetate

It is difficult to make a satisfactory chemical estimate of the molybdenum content of composts when this element is at deficiency levels; plant diagnosis is usually considered to be a better method.

The micro-element content of peat bogs has been studied in Scotland by Mitchell (1954) and in Ireland by Walsh and Barry (1958); typical values reported by Mitchell are given in Table 7.1. Walsh and Barry found the micro-element concentrations varied with the depth of the profile. The amount of elements in the top layers was usually greater than at a depth of 3 m, but large variations were found to exist both within and between bogs. They concluded that atmospheric precipitation of elements in sea-derived salts and in dust were important sources of minerals in bog development.

Table 7.1. *Micro-element content of some Scottish peats.*

Peat source	Composition	Sample depth	% Ash	Micro-element (ppm dry wt)				
				Mo	Fe	Zn	Cu	Mn
Red Moss (Aberdeenshire)	Younger sphagnum (Calluna-Eriophorum)	150–165 cm	1·1	0·17	300	6	1·1	4·6
Aird's Moss (Ayrshire)	Sphagnum-Eriophorum	60 cm	0·9	0·23	175	10	1·7	1·6
Westerdale (Caithness)	Sphagnum-Scirpus	15–45 cm	2·3	0·60	580	10	6·3	2·5
Loch Chalium (Caithness)	Sphagnum-Scirpus	15–45 cm	2·1	0·50	1120	8	13·2	4·2

7.1 BORON

Boron is associated with several functions in the plant, e.g. respiration, carbohydrate metabolism, the transport of sugars and the absorption of calcium. It is also essential for normal cell division and differentiation. Boron deficiency frequently causes the death or blindness of the growing point, resulting in the production of numerous axillary shoots thereby giving a 'witches broom' effect. Deficiency symptoms vary somewhat with the plant species; in tomato, the leaves show a yellow chlorosis, sometimes with reddish-purple pigmentation and often accompanied by die-back from the tips; they are fleshy, brittle and have a high sugar content, and they may also show a red-brown colour in the veins when viewed from the underside. The fruit usually shows a ring of corky splits around the calyx. In the cucumber, the deficiency is shown as cupping of the leaves and death of the growing-point. Cauliflower leaves are much reduced in width, somewhat resembling molybdenum deficiency, and are puckered, whilst, in celery, numerous latitudinal splits or cracks appear on the undersides of the stems. In chrysanthemum, the florets or petals are in-curved and quill-like; this is most noticeable in the broad petal reflexed flower varieties such as the Princess Anne group of varieties as shown in Figures 7.1 and 7.2. In the carnation, the deficiency is frequently seen as the 'witches broom' effect. Antirrhinum shows puckering of the leaves with yellow-to-gold chlorotic patches; the main shoot is often blind giving the plant a highly branched appearance.

Experiments have shown that plants grown in composts having organic- or ammonium-producing forms of nitrogen in the base fertiliser are more susceptible to boron deficiency than those having nitrate nitrogen (Bunt, 1972). This is caused partly by the rise in the compost pH which occurs as the organic nitrogen is first converted into ammonium before nitrification com-

Fig. 7.1 Boron deficiency in chrysanthemum flowers. Deficient flower on the left shows an incurved form with the florets being quilled. Normal flower on the right.

Fig. 7.2 Normal florets (petals) on the left, the tubular or quilled florets from boron deficient flowers on the right.

mences, and is partly the result of a temporary lock-up of the boron following the big increase in the micro-organism population as the organic nitrogen is mineralised. An example of the effect of the nitrogen source on the incidence of boron deficiency is seen in Figure 7.3. Antirrhinums grown with hoof and horn or urea as the nitrogen source showed typical boron deficiency symptoms, whereas plants grown with calcium nitrate were normal.

Fig. 7.3 Plants grown with ammonium-producing fertilisers are more susceptible to boron deficiency. *Left:* antirrhinum in peat-sand compost with hoof and horn as the nitrogen source. *Centre:* urea as the nitrogen source. *Right:* calcium nitrate as the nitrogen source.

134 *Micro-Elements*

Symptoms of boron toxicity are first seen as a necrosis around the edges of the leaves. This is caused by boron being carried in the transpiration stream and then deposited at the extremity of the leaves where the localised concentration can be several times greater than in the central parts of the leaves; this symptom is shown in the chrysanthemum leaves in Figure 7.4. Plants such as *Beloperone guttata,* chrysanthemum, garden pea, poinsettia and zinnia are very sensitive to boron toxicity, whilst the carnation and china aster are relatively tolerant.

Fig. 7.4 Boron toxicity in chrysanthemum leaves shows as a leaf edge necrosis.

Boron can be added to the compost with the base fertilisers, either as an inorganic salt, such as sodium borate or borax ($Na_2B_4O_7 \cdot 10H_2O$, having 11·3% B), or in a general mixture of micro-elements in a fritted form. With crops that have a high boron requirement, such as carnations grown in glasshouse borders, boron can also be given regularly at 0·5 ppm B in the liquid feed. The limits between plant deficiency and toxicity levels with this element are very narrow. Bunt (1971) found that when chrysanthemums were grown in a peat–sand compost without any added boron, deficiency symptoms were produced in the flowers (Fig. 7.1) and when 22 grams of borax were added per cubic metre of compost (0·6 oz./cu.yd), toxicity symptoms developed in the leaves. A dressing of 7 g/cu.m (0·25 oz./cu.yd) of borax was found to be sufficient to correct the deficiency without causing toxicity symptoms. Such low rates of application, however, call for very thorough control of the mixing, and a much safer method of applying boron to composts is to use the fritted form. This can be regarded as being a 'slow-release' form of boron with a low risk of toxicity; further discussion on the use of microelements in fritted form is given on page 143.

Boron can be safely used in liquid fertilisers only where an accurate and reliable means of diluting the stock solution is available, otherwise toxicity could soon occur. If boron is to be included in the liquid fertiliser, a form known as 'Solubor' can be used. This has a formula of $Na_2B_8O_{13}.4H_2O$ with 20·5% B, i.e. it has almost double the boron content of borax, and its cold water solubility is about three times that of borax. *It is very important to remember which of these forms of boron is being used when calculating the amount required to make the stock solution.* The quantities of either of these two materials, required to give 0·5 ppm of boron in the liquid feed, are given in Table 7.2. It is assumed that a stock solution is first made and then diluted at the rate of 1 in 200. For *regular* or *frequent* applications, the strength should not exceed 0·5 ppm B; where only an occasional application of boron is required to correct a deficiency, the strength can be increased to 2 ppm.

Table 7.2. *Weights of boron salts required to prepare stock solutions to give 0·5 ppm of boron after dilution at 1 part in 200.*

Stock solution, to be diluted 1 in 200.	Sodium borate $(Na_2B_4O_7.10H_2O)$ (11·3% B)	'Solubor' $(Na_2B_8O_{13}.4H_2O)$ (20·5% B)
Grams per litre	0·885	0·488
Ounces per 10 Imperial galls	1·4	0·78
Ounces per 10 USA galls	1·2	0·65

Alternatively, boron can be applied as a foliar spray to plants showing deficiency symptoms, a 0·2% solution of solubor, i.e. 0·2 kg in 100 l or 0·2 lb. in 10 galls, gives 410 ppm of boron, *this solution must not be applied direct to the compost,* and excessive run-off into the pots should be avoided.

7.2 COPPER

Copper is involved in respiration and oxidation reduction reactions in plants and it functions in several of the enzyme systems. Plants deficient in copper are low in ascorbic acid oxidase, a copper-containing enzyme. Copper is also necessary for the utilisation of iron. In certain cases (Bunt, 1971) chlorosis in chrysanthemum leaves has been controlled by the application of either copper or iron separately.

Deficiency symptoms in plants vary considerably. In fruit trees, it is seen as a rosetting of the terminal leaves; in lettuce plants there is failure to form hearts, with the leaves being narrow and cupped. Tomatoes show inward rolling of the leaves and stunted growth. With chrysanthemum, deficient plants have narrow leaves with interveinal chlorosis and the lower axillary shoots fail to develop. Flowering is delayed by 10 days in cases of mild deficiency, with a complete failure of the flowers to open in severe cases. Plants grown with high levels of nitrogen have a higher requirement for copper, expecially when the nitrogen is in the ammonium form.

Copper deficiency frequently occurs in peat and organic soils, especially when the pH is below 4·5 and the level of extractable copper is below 3 mg/l. of copper. The deficiency can be corrected by applying 10–20 g of copper sulphate per cubic metre. Raising the pH value of very acid peats to 5·5 increases the availability of the copper, but this decreases again above pH 6·0; the natural chelates formed by copper in peat also increase its availability to plants. Foliar sprays of 0·2% copper oxychloride (50% Cu) or cuprous oxide are more effective than copper sulphate, which in some instances can cause scorching of the leaves.

Copper is another of the micro-elements which can cause phytotoxicity at high concentrations, toxicities having been reported in citrus and gladiolus. Often the roots of plants grown in composts having very high levels of copper may contain up to ten times as much copper as the shoots. Excessive copper uptake usually reduces the iron content of the leaves and applications of iron chelate and phosphorus to the compost will reduce the effects of copper toxicity.

7.3 MANGANESE

Manganese is a constituent of certain enzyme systems concerned with respiration, nitrogen metabolism and the transference of phosphate. Whilst it is not a constituent of chlorophyll, it is found in high concentrations in the chlorophyll-containing tissue and a deficiency of manganese prevents chlorophyll formation. Manganese also affects the form in which iron is present in the leaves. When manganese is deficient, the iron is oxidised from the ferrous to the non-available ferric form and in some plants this is deposited in the leaf veins.

Deficiency symptoms may occur as a chlorotic marbling or as specks on either the young or old leaves, and is followd by necrotic spots, the small veins remaining green. Brassica crops are regarded as being particularly sensitive to manganese deficiency. Only the exchangeable and water-soluble forms of manganese are available to plants and, when new peat is limed and the pH raised to above 6·0, the amount of available manganese is very low and plants may soon show deficiency symptoms. The deficiency can be prevented by including with the base fertiliser 5–10 g of manganese sulphate per cubic metre of compost. Deficiency symptoms in plants can be corrected by applying foliar sprays of 0·1% manganese sulphate or manganese in a chelated form.

High levels of available manganese in the compost can be toxic to young seedlings. Symptoms of manganese toxicity in tomato seedlings are first seen in the cotyledons, which turn yellow and die early. Older leaves point downwards and the leaflets curl under; the veins show a dark purplish-brown necrosis with some brown spotting occurring in the adjacent tissue. The young leaves show chlorosis and the seedlings have a generally stunted appearance. Manganese toxicity can occur under three general conditions:

(1) Anaerobic conditions in the compost caused by poor drainage.
(2) Following steam sterilisation of certain types of soil.
(3) Under acid conditions.

Needham (1973) has reported toxicity symptoms in the Princess Anne varieties of chrysanthemum when grown in a peat–sand compost having a pH of 4·9 with an extractable manganese level of 24 mg/l. Leaves of the affected plants contained 1 600 ppm of manganese; the peat used to make the compost had not been steam sterilised. Raising the pH of the compost with lime removes the toxicity by reducing the amount of available manganese.

7.4 MOLYBDENUM

Molybdenum is required by plants for the reduction of nitrate nitrogen to ammonia within the plant before proteins can be formed. The quantity of this element necessary for the normal functioning of the leaves is very low, even by microelement standards; often the molybdenum level in healthy leaves is less than 1 ppm of the dry weight.

The most frequently observed symptoms of molybdenum deficiency are pale coloured leaves followed by marginal scorching. Tomato leaflets show loss of chlorophyll with an upward curling, followed by withering commencing from the apical leaflet. Lettuce is very susceptible to molybdenum deficiency in the seedling stage. The outer leaves turn pale and then die. Probably the best known symptom of molybdenum deficiency is 'whiptail' of cauliflower, which occurs as a severe restriction on the width of the leaf blade; in some cases the leaf may consist of little more than the midrib. Molybdenum is required in such small quantities that plants which are vegetatively propagated, such as chrysanthemum, frequently fail to show deficiency symptoms if the cuttings have been taken from stock plants which were adequately supplied with the element. Certain seed-propagated plants such as lettuce and brassica are very susceptible to the deficiency. Results of an experiment made on the molybdenum requirement of lettuce seedlings (Bunt, 1973, *a*) show that a response was obtained from the addition of 5 g/cu.metre of ammonium molybdate to the seed sowing compost (Table 7.3). A further response in plant growth was obtained when the ammonium molybdate was

Table 7.3. *Response of lettuce, cv. Kwiek, to molybdenum added to the seed sowing and potting compost. (Fresh weight in grams.)*

Seed sown in peat–sand seed sowing compost	Seedlings grown in peat–sand potting compost with Frit 253A at 375 g/cu.m		John Innes Potting compost (JIP-1)
	No extra molybdenum	Extra molybdenum 5 g/cu.m of ammonium molybdate	
No molybdenum	11·2	32·6	26·2
With 5 g/cu.m of ammonium molybdate	23·2	33·1	27·6

Frit 253A contains: 2% B, 2% Cu, 12% Fe, 5% Mn, 0·13% Mo, 4% Zn.

added to the potting compost, even although the potting compost contained a fritted trace element mixture which included some molybdenum. The quantity of molybdenum supplied by this fritted trace element mixture was obviously not sufficient, and subsequent to this work a new formulation of frit with a higher molybdenum content has been made (p. 145).

Primula also appears to be sensitive to molybdenum deficiency; plants grown in winter have been reported as having a greater molybdenum requirement than summer grown plants (Reeker, 1960). Those plants receiving nitrogen in the nitrate form are more likely to show deficiency symptoms than plants receiving the ammonium form. High levels of sulphates in the compost can also reduce the uptake of molybdenum by plants.

The deficiency can be controlled by including with the base fertiliser either ammonium or sodium molybdate at 5–10 g/cu.m of compost; toxicity has been reported at a rate of 100 g/cu.m. Molybdenum can also be used in the fritted form; the mixture of fritted trace elements should contain not less than 1% by weight of molybdenum. Poinsettia can be very susceptible to this deficiency when propagated and grown under certain conditions, and, as a safeguard, molybdenum is often included at 0·1 ppm in the liquid feed. This strength can be obtained by dissolving 5 g of sodium molybdate ($Na_2MoO_4 \cdot 2H_2O$, 39·7% Mo) in 100 l of stock solution and diluting at 1 in 200 (4·6 g per 20 Imperial galls or 3·8 g per 20 USA galls, and dilute at 1 in 200). Molybdenum can also be applied as a 0·05% sodium molybdate spray.

7.5 IRON

Iron deficiency is one of the most common deficiencies seen by horticulturists. Whilst most plants can suffer from this deficiency if conditions are adverse, azalea, calluna, camellia, erica, gerbera, hydrangea, magnolia, meconopsis, petunia, primula and rose are particularly susceptible. Iron is a constituent of certain enzyme systems, e.g. peroxidase and catalase, and, whilst it is not a constituent of chlorophyll, it is necessary for its formation. Iron deficiency is first seen in the young expanding leaves as an interveinal chlorosis. As the symptoms develop, the light green areas turn yellow and in acute cases the leaf may turn white and become completely devoid of chlorophyll. Often the deficiency can be corrected if the treatment is given whilst the leaves are still young and expanding; when the leaves have fully developed correction measures are seldom successful.

The relatively frequent occurrence of iron deficiency is due to its having more than one cause. In addition to physical and chemical factors in the compost, the environment is also known to affect its incidence. Wallace and Lunt (1960), in a comprehensive review of iron deficiency in horticultural plants, list more than ten possible causes, the most important are:

(1) High light intensity.
(2) High or low temperature.
(3) Over-watering and poor aeration.
(4) Unbalanced cation ratios.
(5) Low iron supply.
(6) Calcium carbonate in the soil.
(7) Bicarbonate in the soil or water.
(8) High level of phosphate.
(9) High levels of heavy metals such as manganese copper and zinc.
(10) High level of nitrate nitrogen.

One of the most frequent causes of iron deficiency of pot-grown plants is root damage; this can be caused by specific toxicities, high salinity or by over-watering. One example of iron deficiency resulting from root damage is shown in Figure 7.5. The roots of the chrysanthemum were damaged by high ammonium levels following a heavy application of an organic nitrogenous fertiliser and as a result nine leaves were produced with acute iron deficiency symptoms. After a period of recovery, during which new roots were formed, normal green leaves were produced again, no corrective measures having been taken in the meantime. Excessively wet conditions, caused by over-watering a poorly drained compost, can also produce iron deficiency and it is often beneficial to keep plants in a drier state for a few days after the symptoms first appear. High pH values, associated with over-liming, result in a condition known as lime-induced chlorosis. Apart from its effect upon iron availability, high pH values in peat-based composts should be avoided, because of the adverse effect on the availability of other micro-elements and the greater risk of ammonium toxicity.

140 *Micro-Elements*

Fig. 7.5 Chrysanthemum showing iron deficiency caused by root damage from high ammonium levels. Note the normal green leaves produced before and after the period of root damage.

The main nutritional interactions which are known to have depressing effects upon iron metabolism are: high levels of phosphorus in the compost resulting in the formation of insoluble iron phosphates, high potassium to calcium ratios within the plant and high levels of copper and manganese. Under conditions of low winter light intensity and low carbohydrate supply, growers prefer to apply most of the nitrogen in the nitrate rather than the ammonium form. However, some plants such as the pot-grown chrysanthemum can show incipient iron deficiency symptoms when given too much nitrogen in the nitrate form; some ammonium nitrogen should always be present.

Whilst the soluble iron content of acidic peats is normally greater than that of neutral or alkaline mineral soils, iron deficiency can occur in plants grown in the younger sphagnum peats. This can be corrected by applying ferrous sulphate or an iron chelate at 10 to 20 g/cu.m of compost ($\frac{1}{4}$ to $\frac{1}{2}$ oz./cu.yd) with the base fertiliser; the chelated form of iron will usually give a better result. When deficiency symptoms first appear in the plants, a 0·1% solution of iron chelate applied to the compost will usually be more effective than a foliar spray. Chelated forms of micro-elements are described more fully on page 145. Read and Sheldrake (1966) have reported iron deficiency symptoms in plants grown in peat–vermiculite mixes; this was attributed to the high Mn:Fe ratio in the vermiculite. The addition of large amounts of lime, phosphorus, zinc and manganese and the use of water high in carbonate and bicarbonate ions all increased the chlorosis; foliar sprays of chelated iron eliminated the trouble.

Iron deficiency is unique because affected leaves may contain as much or even greater amounts of iron than healthy leaves. Deficiency symptoms are often due to the unavailability of the iron rather than to a true deficiency. More recently biochemical methods of diagnosing the deficiency by comparing the enzyme activity of healthy and affected leaves have shown promise.

7.6 ZINC

Zinc deficiency is not generally encountered in pot grown plants. It has, however, been reported in Florida in plants grown on organic soils. The deficiency has occurred in citrus and peach trees and is known as 'mottle leaf' and 'little leaf'. A small quantity of zinc sulphate ranging from 0·5 to 5·0 g/cu.m of compost is often included with the base fertiliser as a precautionary measure; where zinc deficiency is suspected a foliar spray of 0·05% zinc sulphate can be applied.

Zinc toxicity in plants can sometimes occur when new galvanised (zinc coated) pipes are used in the water supply system. Tests have shown that the first few gallons of water drawn off in the morning, after water has been static in the pipes overnight, can contain as much as 4 ppm of zinc. This value soon falls as the pipes are flushed with fresh water and the zinc level will then fall to 0·5 ppm or less. When the pipes have aged, the zinc value will often be not greater than 0·1 ppm. Static water drawn from untreated galvanised tanks can contain as much as 10–20 ppm of zinc. In water culture studies it has been found that the zinc level should not exceed 0·2 ppm. Localised zinc toxicity of plants can also occur from condensation dripping from newly galvanised wire and tubular blackout supports such as are used in chrysanthemum growing.

7.7 CHLORIDE

The requirement of plants for chloride is very low and it can be shown to be an essential element only by growing plants in water cultures and working under closely controlled conditions. Chlorides usually occur in sufficient quantity in the irrigation water and as impurities in the fertilisers to supply all the plant's requirements. Normally, chlorides are of concern only when they reach toxicity levels. This shows as a burning or scorching of the leaves, together with bronzing and premature defoliation. In some parts of the world, chlorides are present in the irrigation water at sufficiently high concentrations to cause toxicity, i.e. >70 ppm Cl. In England, the chloride content of the public water supply usually does not exceed 20 ppm. Similarly with free chlorine, public water supplies do not normally contain more than 0·5 ppm of free chlorine. Water containing 2 ppm of free chlorine is not drinkable and plants have been given water with 5 ppm of free chlorine without showing any toxic symptoms.

American workers have examined the reaction of a number of ornamental plants to chlorides and found gardenia, geranium and poinsettia to be particularly sensitive. Stocks were classified as being intermediate and carnation, penstemon and verbena were regarded as being tolerant to chlorides. Arnold Bik (1970), in the Netherlands, examined the effect of chloride and nitrogen levels on the growth and chemical composition of gloxinia and chrysanthemum. He found the detrimental effect of chloride was primarily due to the related increase in the salinity as measured by the electrical conductivity of a saturated extract of the compost, i.e. the EC_e; there was also some evidence of a specific ion effect. Gloxinia, which is a salt sensitive plant, showed a decline in growth when the EC_e rose above 4 millimhos, while for the chrysanthemum, which is a more tolerant plant, the corresponding value was 6–8 millimhos.

7.8 ALUMINIUM

This element is not essential to plant life and normally receives attention only when a toxicity arises, e.g. in plants grown in very acid soils. Many plants accumulate aluminium, e.g. the tea plant.

The hydrangea is of interest because the colour of its flowers is dependent upon the amount of aluminium they contain. In its native habitat this plant produces flowers (sepals) which are naturally blue. In some locations, however, the flowers are often pink. This phenomenon was investigated by Allen (1943), who grew plants in soils with pH values ranging from 4·6 to 7·4, and related the pH of the soil to the colour and aluminium contents of the flowers; his results are summarised in Table 7.4. By the use of foliar sprays and the absorption of salts through cut stems, he showed that the flower colour was dependent upon the amount of aluminium in the tissue. Iron salts had no direct effect upon the colour but they had an indirect effect by reducing the pH, which in turn increased the availability of the aluminium. Iron salts are, however, beneficial in preventing iron chlorosis, a deficiency to which this plant is particularly susceptible.

To produce blue hydrangea flowers it is customary to grow the plants in a compost with a pH of 4·5–4·8; above this range there is a marked re-

Table 7.4. *Effect of soil pH on the colour and aluminium content of hydrangea flowers.*

pH Range	Colour	Aluminium (ppm)
4·6–5·1	Blue	2 375–897
5·5–6·4	Blue, tinged pink	338–187
6·5–6·7	Pink, tinged blue	214–201
6·9–7·4	Clear pink	< 217

duction in the amount of aluminium in solution. Normally, 14 kg/cu.m (5 lb./cu.yd) of aluminium sulphate is included with the base fertiliser; this helps to maintain a low pH and also provides sufficient soluble aluminium. When the plants are brought into the glasshouse in January for forcing, after having had a period of low temperatures to break their bud dormancy, a 1% solution of aluminium sulphate is given at 14 day intervals. High levels of phosphorus in the compost will result in the formation of insoluble aluminium phosphates and for this reason plants which are to be 'blued' are given only low levels of phosphorus. When pink flowers are required, the soluble aluminium must be kept at a low level by growing at pH 6·5, omitting all aluminium sulphate treatments and maintaining high levels of phosphorus.

7.9 FRITTED MICRO-ELEMENTS

One problem encountered in supplying micro-elements to composts is the very narrow limits which exist between deficiency and toxicity levels; as previously shown, boron toxicity occurred when sodium borate was used at rates above 7 g/cu.m. Applying such small quantities of a chemical presents problems in the weighing and uniform mixing into the compost. Boron is also quite soluble in organic media such as peat and when plants require frequent watering there is some risk of its loss by leaching, with a deficiency occurring later on. One method of overcoming this problem is to supply boron in a slowly available form such as a soft glass frit; in this way the safety margin is increased and the risk of leaching decreased. Glass frits are used in industry in the manufacture of porcelain enamel and ceramic glaze finishes. For agricultural purposes, frits, which are known as FTE, i.e. fritted trace elements, are prepared by adding the desired amounts of inorganic salts to sodium silicate. This is heated to a temperature of around 1 000°C and the molten mass is poured on to either cold steel or running water, where it cools and shatters into small fragments which are ground into a very fine powder. The rate at which the frits dissolve in the compost and release their micro-elements to plants can be controlled by adjusting the rate of cooling, the fineness of the grinding and the addition of chemical impurities. Some of the properties of frits have been reviewed by Holden, Page and Wear (1962).

144 Micro-Elements

Frits were considered a satisfactory method of supplying boron to plants but they are less efficient in supplying iron.

Once the fritted trace element mixture has been prepared, the rate at which minerals are released is primarily controlled by the pH of the compost. Bunt (1972) examined the effects of the pH of the compost and the rate of application of a fritted trace element mixture (FTE 253A) containing 2% B, 2% Cu, 12% Fe, 5% Mn, 0·13% Mo, 4% Zn, on the growth and chemical composition of cauliflower, celery, chrysanthemum, tomato and pea. Large interactions were found among the plant species, the pH of the compost and the rate of frit. Tomatoes grown in composts to which no frit had been added, showed a marked response to the pH, the optimal pH being 6·4 (M/100 $CaCl_2$) with marked depressions in growth occurring above and below this value, Figure 7.6. When the frit was added at the rate of 350 g/cu.m (10 oz./cu.yd), the effect of the compost pH was largely eliminated and also plant growth increased significantly; there was very little difference in plant growth between the 350 g/cu.m and 1·75 kg/cu.m rate (3 lb./cu.yd). With chrysanthemum, however, the 1·75 kg/cu.m rate suppressed growth. This

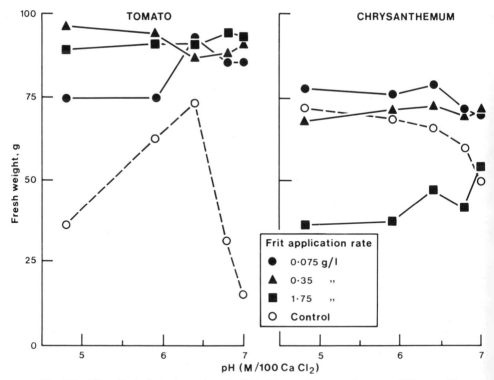

Fig. 7.6 The effect of varying rates of a fritted trace element mixture at different pH values on the growth of tomato and chrysanthemum.

was caused by the amount of boron released from the frit, chrysanthemum being much more sensitive to boron toxicity than the tomato.

As a result of this work and also the experiment on the molybdenum requirement of lettuce (p. 138), a new frit, known as WM 255 and having the following composition, has been formulated (Bunt, 1974, b):

1% B, 4·3% Cu, 13·8% Fe, 5·4% Mn, 1% Mo, 4·3% Zn.

This frit has a lower boron content and also greater amounts of copper and molybdenum than the 253A frit. It has been used successfully at 350 g/cu.m (10 oz./cu.yd) for raising bedding plants, chrysanthemums, poinsettias, lettuce, tomato and other pot plants. Where there is likely to be a very heavy demand for boron, e.g. if tomato plants are cropped with several trusses of fruit in a 25 cm (10 in.) pot, it may be necessary to give an occasional boron feed as the plants develop.

Smilde (1971) found a multi-nutrient frit to be as effective as straight fertilisers in supplying copper, boron and molybdenum to chrysanthemums grown in a 'black' peat, i.e. a humified old sphagnum peat formed from *Eriophorum, Carex* and *Phragmites*.

7.10 CHELATED MICRO-ELEMENTS

Often micro-element deficiencies in plants are the result of an unavailability of an element or antagonisms between elements, rather than a true deficiency of the element. Iron is an example of an element rendered unavailable to plants because of its chemical precipitation in the compost by other elements. To overcome this problem, certain plant nutrients can be treated in a way which keeps them in an available form; the term used to describe this process is 'chelation' and is derived from the Greek word meaning 'claw'.

To be successful, a chelating material must:

(1) prevent the iron from being precipitated in the soil, e.g. by the action of high pH or phosphorus levels,
(2) prevent its replacement by other metals, e.g. zinc,
(3) be non-phytotoxic,
(4) resist decomposition by soil-organisms.

Examples of naturally occurring chelate materials are citrate and tartrate acids; recently synthetic chelates have been successfully prepared from the polyamino-polycarboxylic acids. One of the earlier materials used with success was EDTA (ethylenediamine tetraacetic acid) and several other chelating agents have been developed since. Some chelates are most effective under acid conditions, whilst others are effective under alkaline conditions; EDTA, for example, is most effective when the pH is below 6·0. Wallace (1962) concluded that Fe EDDHA (ethylenediaminedi (0-hydroxyphenyl)-acetate) was the most successful of the iron chelates for use under a range of soil conditions. It is a very safe and very effective material; no other chelating agent is better under alkaline conditions and it is at least as good as the others in acid soils. It is available under the trade names Chel 138 or Sequestrene 138 and contains approximately 10% iron. For foliar application, however, the EDTA form is usually more effective than EDDHA. In addition to being beneficial to plants under conditions of iron deficiency, chelated iron has been found to be beneficial also when toxicities of copper, zinc and manganese exist. Furthermore, a general stimulation of plant growth following the application of chelating agents to non-deficient plants has been frequently observed. Amongst the reasons suggested for this phenomena are better micro-element balance, stimulatory effects on some enzyme reactions and inactivation of calcium in the cytoplasm. Chelates can be added to the compost with the base fertiliser or by applying a 0·1% solution when the plants are growing in the pots; the latter method of application is often more successful than foliar sprays, which should never be applied at strengths above 0·1% to young leaves not yet fully expanded.

Whilst iron is the most widely used of the chelated metals, other micro-elements such as zinc, copper and manganese can be chelated also, and general mixtures of micro-elements in EDTA form are available. Neither boron nor molybdenum can be chelated, however, and they can be included in the mixture only as inorganic salts, usually as sodium borate and ammonium or sodium molybdate; they can, of course, be applied separately in fritted form. It is important to remember that frits are a way of supplying micro-elements in a slowly available form, whereas chelated micro-elements have an immediate availability.

Experiments at the Glasshouse Crops Research Institute with a micro-element mixture based on an EDTA chelate and having an analysis of:

$$1·7\% \text{ Cu}, 3·35\% \text{ Fe}, 1·7\% \text{ Mn}, 0·6\% \text{ Zn}, 0·875\% \text{ B}, 0·023\% \text{ Mo},$$

showed the optimal rate of application to be low; it also varied with the plant species. For zinnia, the optimal rate was 75 g/cu.m (2 oz./cu.yd) whilst for aubergine and exacum it was 150 g/cu.m (4 oz./cu.yd). At double this rate, there was a suppression of growth, the fresh weights being approximately one sixth that of the plants having the optimal rate.

7.11 OTHER SOURCES

Apart from the sources of micro-elements already discussed, indirect sources of supply are impurities in the water, e.g. zinc, and fungicide sprays, e.g. copper. The most important single source of supply, however, is the impurities present in the commercial grades of fertilisers, of which the phosphorus

Table 7.5. *The micro-element content of fertilisers (ppm).*

Fertiliser	B	Cu	Mn	Mo	Zn
Ammonium sulphate	0·2–25	0–20	Tr–80	Tr–0·2	0–100
Ammonium nitrate	0·4–2·0	Tr–1	<5	0·1–0·3	1–5
Urea	0–10	0–4	1–10		0–50
Calcium nitrate	Tr–90	1–20	1–10		<1·0
Sodium nitrate	50–300	1–25	<1	0·1	1–10
Superphosphate (single)	3–15	10–60	10–200	Tr–10	70–500
Superphosphate (triple)	Tr–200	30–200	0–200	3–20	0–100
Monoammonium phosphate	10–100	10–100	30–200	2–10	30–200
Phosphoric acid	<6	15–100	40–2 000	100	1–300
Phosphate rock	<50	1–30	10–200	Tr–20	5–300
Basic slag	20–1 000	10–60	1 000–50 000	Tr–10	3–30
Potassium nitrate	1–2	Tr–30	Tr–8		<8
Potassium sulphate	<30	1–10	Tr–50	0·1–0·3	0–6
Potassium chloride	0–150	0–10	Tr–8	Tr–0·2	<3
Limestone	2–50	2–200	10–700	2–20	3–200
Calcium carbonate	<0·3	0–50			3–30
Dolomite limestone	1–25	1–100	10–500	1–20	5–100

These data indicate only the range of values that are commonly present in commercial fertilisers. The actual amounts of trace elements present depends upon the method of fertiliser manufacture, e.g. whether it is made by a synthetic process or by treatment of raw fertiliser materials, such as rock phosphates; in the latter case the trace element concentration also depends upon the source of the phosphate. Rock phosphates obtained from American sources appear to contain more micro-elements than those from N. Africa.

fertilisers are the most heavily contaminated. Some indication of the amounts of micro-elements present in commercial fertilisers (Swaine, 1962) is given in Table 7.5. The micro-element values given in the table are for a typical range of concentrations found in the fertilisers; because of the very large variation possible between samples no attempt has been made to give either the highest or mean values. It is apparent from the data that it would

Table 7.6. *Average micro-element content of phosphorus fertilisers.*

Fertiliser	Boron	Copper	Manganese	Molybdenum	Zinc
Phosphoric acid	25	62	331	20	989
Superphosphate (Single)	57	70	48	18	1 000
Superphosphate (Triple)	60	70	160	20	1 500
Ammonium phosphate	51	56	149	14	1 054

be impossible to estimate the quantity of micro-elements being supplied by the fertilisers unless an analysis of the actual materials was made. Some average values of the micro-element content of phosphorus fertilisers, derived from the data of Bingham (1959) working in the USA, are given in Table 7.6.

7.12 MICRO-ELEMENT AVAILABILITY

Apart from those occasions when there is a true deficiency of micro elements in the compost, deficiency symptoms in plants can be the result of either:

(a) incorrect pH,
(b) imbalance or antagonism with other micro-elements,
(c) antagonism with macro-elements,
(d) compost too wet or too dry.

Several antagonisms between elements are known to occur and the most important factors controlling the availability of micro-elements are given in

Table 7.7. *Conditions affecting availability of micro-elements in composts.*

Element	Conditions affecting its availability
Boron	Availability reduced at pH >6.8, relatively easily leached from organic matter, e.g. peat. Organic nitrogenous fertilisers reduce availability, by temporary biological lockup and by rise in compost pH during first stages of mineralisation. High phosphorus levels reduce uptake by plants grown in acid media.
Manganese	Availability much increased at low pH. Plant uptake reduced by high K levels and also high levels of Fe, Cu and Zn. Phosphorus application helpful in reducing plant uptake in cases of toxicity following steam sterilisation, providing that pH is not thereby reduced. Fe chelate also useful in cases of toxicity. Ammonium ions have favourable effect in alleviating Mn toxicity in beans and flax.
Copper	High rates of nitrogen, especially in NH_4 form, increases Cu deficiency, also high levels of P. Cu toxicity reduced by Fe chelate.
Iron	Deficiency induced by high levels of Cu, Mn and Zn. High levels of P result in insoluble Fe phosphates. 'Lime induced' chlorosis caused by over-liming and irrigation water high in carbonates and bicarbonates.
Molybdenum	Availability decreased at pH <6.0, also by high levels of available Mn, and by SO_4 ions and copper. Plants receiving NO_3 form of N have greater requirement than those having NH_4-N.
Zinc	Availability reduced at high pH and by applying magnesium carbonate, also high levels of phosphorus reduce availability and plant uptake.

Table 7.7. It will be seen that phosphorus can have a significant effect upon the availability of several micro-elements. Bingham and Garber (1960) compared the effects of several types of phosphorus fertilisers upon micro-element availability and obtained essentially the same results with all of the forms of phosphorus tested.

MICRO-ELEMENTS IN PLANT TISSUE

Advisory Officers now find that a micro-element analysis of the leaves of plants assists in cases where a micro-element deficiency is suspected; the analysis also provides information upon which to base nutritional programmes. Plant species, however, show considerable differences in the concentrations of micro-elements present in their leaves and also in the levels at which deficiencies and toxicities become apparent. For example, *Beloperone* and poinsettia are very much more sensitive to high levels of boron than the carnation. The micro-element values given in Table 7.8 must, therefore, be regarded only as guide lines. As with the macro-elements, the concentration of micro-elements often varies with the position of the leaf on the plant, and a standard procedure in collecting the tissue must be followed.

Table 7.8. *Average micro-element ranges found in the leaf tissue of ornamental plants (ppm in dry matter).*

	Plant condition		
Element	Deficient	Normal	Excessive
Boron	< 20	30–80	> 150
Manganese	< 20	30–200	> 1 000[2]
Copper	< 5	10–25	> 70
Molybdenum	< 0·1	0·1–3	> 70[1]
Zinc	< 15	30–100	> 200

[1] There is very little data available on molybdenum toxicity, 20 ppm is regarded as excessive in cucumbers.
[2] Manganese toxicity occurs in glasshouse grown lettuce at 500 ppm Mn in the dry tissue.

Chapter 8

Compost formulation and preparation

8.1 HISTORICAL

A brief historical review of the development of loamless composts shows the difficulty of attempting to establish a date for their first introduction or general acceptance into horticulture. As with many horticultural practices, evolution has occurred over a long period. Peat, leafmould and pine needles have been used for growing azaleas by many generations of gardeners and experiments made at Versailles in 1892 on the nutrition of azaleas in these media are described by Watson (1913). Laurie in Ohio, USA, experimented in the late 1920s and early 1930s on the use of peat-sand mixtures for growing a range of plants, but no further development of this concept appears to have occurred for the next 20 years. In the 1950s, American workers at Michigan (Asen and Wilding, 1953) and in California (Baker, 1957) revived the interest in the use of peat and sand by obtaining favourable results in comparison with plants grown in traditional composts based on mineral soils. In Europe, Penningsfeld, working in Bavaria, initiated work on the use of pure peat as a compost (Penningsfeld, 1962), and Puustjärvi (1969), working in Finland, developed a system of growing in peat known as 'basin culture'. With this system, vegetable and ornamental crops are grown in peat, isolated from the glasshouse border soil by a sheet of polythene. Several attempts have been made to include clay, in either powdered or granular form, into composts made from peat and sand. Fruhstorfer (1952) used a 1:1 mixture of granulated clay and peat to formulate his 'Einheitserde' compost. Dempster (1958) recommended 10% by volume of clay to 40% peat and 50% sand, whilst Danhardt and Kukle (1959) reported a progressive increase in growth as the amount of clay mixed with the peat was decreased to 10%; the inferior result of the all-peat compost was attributed to the very low pH of the compost in the absence of some clay. Difficulties over supply and quality control have restricted the use of clay in composts; early unpublished work by the author showed that adverse effects on pH, micro-element availability and forms of nitrogen could occur if the clay contained more than a very small amount of calcium carbonate.

During the last two decades, there has been a large increase in the interest shown by research workers and growers alike in loamless composts; the principal reasons for this were discussed in Chapter 1. Each issue of the abstracting journals covering the world's literature on horticulture and plant nutrition contains several references to the use of loamless composts. These often differ considerably in their formulation, because of the wide variations in the characteristics of the materials used and differences in the nutritional requirements of various crops. For example, polystyrene does not contain any plant nutrients, whereas vermiculite is rich in magnesium and potassium; sedge peats have more available nitrogen than sphagnum peats, etc. Also, a few plant species show distinct nutritional preferences, e.g. azaleas prefer NH_4 to NO_3 nitrogen, ericaceous plants prefer a lower pH than most other plants, whilst tomatoes require a high potassium level. Climàtic conditions also alter the optional balance and concentration of nutrients.

In early work by the author on the nutrition of plants in peat-sand composts, large scale factorial experiments were used to study the simultaneous variation in the concentrations of a number of nutrients. For example, a 3^4 design giving a total of 81 treatments was used to study 3 levels each of N, P, K and Ca. A 2^6 design was used to study two levels of six factors, e.g. 3 factors being physical, such as ratios of peat to sand, grades of peat and grades of sand, with the other 3 factors being nutritional, such as amounts of micro-elements, sources of phosphorus and sources of nitrogen. This made a total of 64 treatments. These experiments were repeated in summer and winter to include the effect of the season, and it soon became clear that no one formulation was optimal for all plants under all conditions.

The very large number of published formulae precludes any attempt at compiling a comprehensive list, rather a selection has been made of the more widely known pot plant composts used in various countries.

8.2 DENMARK

A potting compost known as *Garta Jord* is marketed by the Co-operative Horticultural Supply Company. It is based on a mixture of dark peat, straw that has been composted with added nitrogen, and 'Grodan', a proprietary product made from a glasswool fibre produced from basalt rock and limestone. Its formulation is:

> 50% dark peat
> 25% composted straw
> 25% Blue Grodan

Compost Formulation and Preparation

To this is added:

per cubic metre		per cubic yard
9 kg	Ground limestone	15 lb. 3 oz.
400 g	Superphosphate	10 oz.
200 g	Potassium sulphate	5 oz.
60 g	Magnesium sulphate	2 oz.
50 g	Iron sulphate	1 oz.

plus 16 g copper sulphate, 10 g manganese sulphate, 8 g zinc sulphate, 8 g borax and 2 g sodium molybdate.

This supplies the following amounts of macro-nutrients:

N	P	K
—	32	88 mg/l

No nitrogen is included in the base fertiliser; some nitrogen has already been added to assist in the decomposition of the straw and will be present in the compost largely in organic form. Liquid feeding with 0·2–0·5 g/l of calcium or potassium nitrate is given shortly after potting.

The German 'Einheitserde T' potting compost, based on 50% of an acid montmorillonite clay and 50% peat is also widely used in Denmark.

8.3 FINLAND

Professor Puustjärvi of the Peat Research Institute at Hyrylä has worked extensively on peat as a growing medium for bed or border grown crops such as tomato, carnation, etc., and has introduced the 'basin systems' of cultivation. This consists of sheets of polythene spread on top of the glasshouse soil; the polythene is turned up 1–2 cm at the sides and edges, and peat is placed on top to a depth of 10–30 cm, depending upon the type of crop. The polythene forms a shallow basin and prevents extensive loss of nutrients by leaching. For growing plants in pots, the following nutrient additions per cubic metre of sphagnum peat are recommended (Puustjärvi, 1973):

Dolomite limestone 8 kg
Compound fertiliser 1–1·6 kg (1 lb. 11 oz.–2 lb. 11 oz./cu.yd)

depending upon the vigour of the plant.

The fertiliser has an analysis of 11% N, 24% P_2O_5, 22% K_2O; at 1 kg/cu.m the nutrients supplied would be:

N	P	K
110	105	182 mg/l

In addition, micro-elements included in the fertiliser would supply:

B	Cu	Mn	Zn	Fe	Mo
1	11	9	7	9	1 mg/l.

In the original formulation, several straight or single nutrient fertilisers were used, but a compound fertiliser is now preferred for simplicity. Normally, slow-release forms of nitrogen are not used; liquid feeding with frequent chemical analysis to check the level of plant nutrients is preferred.

8.4 GERMANY

Professor Penningsfeld from the Institute for Soil Science and Plant Nutrition, Weihenstephan, Nr Munich, has made large contributions to our knowledge of the nutrition of ornamental plants grown in peat. He has recommended three rates of nutrients, depending upon the vigour of the plants.

For all composts, calcium carbonate is added at 2–5 kg/cu.m of peat, the actual rate depending upon the lime requirement of the peat and the desired pH. A compound fertiliser having an analysis of 12% N, 12% P_2O_5, 16% K_2O, 2% MgO is then added at the following rates:

Compost 1. 0·5 to 1·0 kg/cu.m, giving the following concentrations of applied nutrients:

N	P	K
60–120	26–53	66–133 mg/l

This compost is used for growing salt-sensitive plants: *Adiantum, Erica gracilis, Primula, Gardenia, Camellia* and certain bedding plants, i.e. *Begonia, Verbena, Godetia, Callistephus* and *Dianthus*.

Compost 2. 1·5 kg/cu.m of compound fertiliser, giving:

N	P	K
180	80	200 mg/l

This compost is used for growing moderately salt-tolerant plants: *Aechmea, Vriesia, Freesia, Anthurium, Gerbera, Aphelandra, Cyclamen, Monstera, Sansevieria, Rosa, Hydrangea*, and certain bedding plants, i.e. *Salpiglossis, Tagetes, Zinnia, Matricaria, Penstemon, Dianthus, Campanula*.

Compost 3. 3 kg/cu.m of the compound fertiliser, giving:

N	P	K
360	160	400 mg/l.

G

154 Compost Formulation and Preparation

This compost is used for growing plants that have the greatest salt tolerance:

Pelargonium, Euphorbia, Saintpaulia, Chrysanthemum, Asparagus and *Carnation.*

Micro-elements are added to these composts in either liquid or metal alloy formulations. An example of a fertiliser supplying micro-elements in liquid form is 'Micro T', made by Gabi, Hundersen, 4902 Bad Salzuflen 1. This contains 0·02% boron, 0·13% copper, 0·05% molybdenum, 0·17% chelated iron, 0·05% manganese and 0·007% zinc and is applied at the rate of 1 l/cu.m. The micro-elements in this fertiliser are immediately available to plants. An alternative fertiliser used to supply micro-elements in a slow release form is 'Radigen', manufactured by Metalldunger Jost, 5860 Iserlohn, Baarstrasse 122. This fertiliser contains 2% magnesium oxide, 2% iron, 1·5% copper, 0·8% molybdenum, 0·8% manganese, 0·8% boron and 0·6% zinc. It is based on metal alloys which are slowly attacked by the weak acids in the compost, thereby releasing the micro-elements over a long period, and is used at the rate of 100 g/cu.m.

8.5 IRELAND

Workers at the Agricultural Institute, Kinsealy, Dublin, have developed a compost based on peat which is suitable for a wide range of plants and is known as the 'Range Mix'. The following fertilisers are added:

per cubic metre		per cubic yard
1·4 kg	Calcium ammonium nitrate (26%N)	2 lb. 4 oz.
0·7 kg	Superphosphate	1 lb. 2 oz.
0·7 kg	Potassium sulphate	1 lb. 2 oz.
2·2 kg	Kieserite	3 lb. 12 oz.
5·6 kg	Ground limestone	9 lb. 4 oz.

This gives an added nutrient concentration of:

N	P	K
364	55	310 mg/l

The following micro-elements are also added (g/cu.m):

Borax	9·4	Zinc sulphate	14·2
Copper sulphate	14·2	Sodium molybdate	2·4
Ferrous sulphate	42·5	Chelated iron	23·4
Manganese sulphate	14·2		

A mix designed specifically for tomato propagation contains the same amounts of ground limestone, Kieserite and potassium sulphate as the Range Mix, but the superphosphate is increased to 1·4 kg/cu.m (2 lb. 4 oz./cu.yd), the calcium ammonium nitrate is reduced to 0·7 kg/cu.m (18 oz./cu.yd) and urea formaldehyde added at 0·7 kg/cu.m (18 oz./cu.yd) to give some slow-release form of nitrogen.

8.6 NETHERLANDS

A standardised potting compost was introduced in the Netherlands in 1964 by the Aalsmeer Research Station for Floriculture (Arnold Bik, 1973). It is known as the Regeling Handelspotgrand Proefstation Aalsmeer, abreviated to RHPA (Regulation System for Commerical Potting Composts Research Station Aalsmeer). This is prepared from a mixture of:

    10 parts by volume sphagnum peat,
    10 parts by volume frosted decomposed black sphagnum peat,
    1 part by volume river sand.

To one cubic metre of this mixture is added 7 kg magnesium limestone (5% MgO) and 1·5 kg of a compound fertiliser having 16% N, 10% P_2O_5, 20% K_2O, plus 150 grams of triple superphosphate and 250 grams of 'Sporumix PG', a commercial fertiliser containing 25% MgO, 0·3% Cu, 0·1% B, 0·5% Mn, 0·15% Zn and 0·6% Mo. This gives the following rates of added nutrients:

N	P	K	B	Cu	Mn	Zn	Mo
240	92	249	0·25	0·75	1·25	0·375	1·5 mg/l

The mixture has a pH (determined in a water suspension) of between 5·5 and 6·5, and is used for most pot plants except those favouring a lower pH, e.g. calceolaria, primula, begonia, cineraria, brunfelsia, etc. For these plants, a calceolaria RHPA mix was introduced in 1970. This is similar to the general RHPA mix, apart from a reduction in the magnesium limestone to 3·5 kg/cu.m to give a pH of 4·6–5·4, and the addition of 25g/cu.m of Chel 138 Fe, an iron sequestrene. This mixture is not, however, recommended for azaleas; the pH and the nutrient levels are considered to be too high.

8.7 UNITED KINGDOM

Various nutritional problems arising from the use of peat-based composts have been investigated by the author at the Glasshouse Crops Research Institute. Special attention has been given to nitrogen and micro-element nutrition, the problems which occur most frequently. Results from this work have been used to formulate seed sowing and potting composts.

GCRI Seed compost

                50% by volume sphagnum peat
                50% by volume fine, lime-free sand

To this is added:

per cubic metre		per cubic yard
0·75 kg	Superphosphate	1 lb. 4 oz.
0·4 kg	Potassium nitrate	10 oz.
3·0 kg	Ground limestone	5 lb. 4 oz.

The seed sowing compost contains a higher proportion of sand than the potting compost; this makes transplanting or pricking-out of young seedlings easier and faster than from a compost with a low sand content. This compost is low in nutrients and is suitable only for seed germination; seedlings should be pricked-out as soon as possible after germination, otherwise they will starve. If for some reason pricking-out is delayed, a solution of calcium nitrate at 1·5 g/l (1·0 oz./5 galls) will maintain the growth of the seedlings until they can be handled. A fritted trace element mixture, WM255, used at 375 g/cu.m when sowing lettuce seeds will prevent molybdenum deficiency arising in the seedling stage. Normally, micro-elements are not required in seed sowing composts.

GCRI Potting Compost I. A general purpose potting compost is made from:

75% by volume sphagnum peat
25% by volume fine, lime-free sand.

To this mixture is added:

per cubic metre		per cubic yard
0·4 kg	Ammonium nitrate	10 oz.
1·5 kg	Superphosphate	2 lb. 10 oz.
0·75 kg	Potassium nitrate	1 lb. 5 oz.
2·25 kg	Ground limestone	4 lb.
2·25 kg	Dolomite limestone	4 lb.
0·375 kg	Fritted trace elements (WM 255)	10 oz.

This gives an added nutrient content of:

N	P	K
230*	120	290 mg/l

*Approximately 30% of the nitrogen is in the NH_4 form and 70% in the NO_3 form.

This compost can be used immediately or stored for an indefinite period before use.

GCRI Potting Compost II. If a compost with a slow-release form of nitrogen is required, the ammonium nitrate in the above formulation can be omitted and urea-formaldehyde used at rates ranging from 0·5–1·0 kg/cu.m (1–2 lb./cu.yd) depending upon the vigour of the species and the season. Slow growing plants, in winter, would receive the low rate and vigorous plants, in summer, the high rate.

This compost must not be stored before use as ammonium toxicity can develop during storage.

GCRI Potting Compost III. If there is the risk of phosphorus deficiency developing in the compost and phosphorus cannot be included in the liquid feed, the superphosphate in Potting Compost I can be omitted, the ammonium nitrate reduced to 0·2 kg/cu.m (5 oz./cu.yd), the potassium nitrate reduced to 0·4 kg/cu.m (10 oz./cu.yd) and one of the magnesium ammonium phosphate fertilisers, such as 'Mag Amp' or 'Enmag' used at the rate of 1·5 kg/cu.m (2 lb. 8 oz./cu.yd). This compost supplies sufficient phosphorus to grow a 10 week pot chrysanthemum crop without the need to include phosphorus in the liquid feed. The phosphorus supply would not, however, be adequate for plants grown for a longer period, e.g. cyclamen.

In all the above composts, the quantity of ground limestone and Dolomite limestone is based upon the lime requirement of Irish Sphagnum peat. If peats with a different lime requirement are used (p. 73), the amount of liming materials must be varied accordingly. For *Ericas*, the pH of the compost should be 4·0–4·5, measured in a suspension with $M/100$ $CaCl_2$. The possible rise in the pH of the compost due to 'hard' water must be borne in mind also. Plants such as cinerarias, calceolarias and primulas have been found to grow better when the pH of the compost is reduced to 4·5 to 5·0.

Early experiments had shown the need to supply micro-elements in the base fertiliser and small amounts of inorganic salts were used. Owing to the toxicity of borates at even relatively low concentrations, and the difficulty of achieving a uniform mix, it is preferable to add the micro-elements in the very much safer 'fritted' form (p. 143). In the initial work with fritted trace elements, Frit 253A, having the following composition, was successfully used for a large number of species:

	B	Cu	Fe	Mn	Zn	Mo
Frit 253A	2·0	2·0	12·25	4·9	4·0	0·13%

Subsequent work showed that the molybdenum content of this frit was not sufficient for lettuce and brassica plants, both species being particularly susceptible to this deficiency. Also the boron content of the frit was found to be rather high for chrysanthemum and poinsettia, two plants which are very susceptible to boron toxicity if the compost pH is low. A new frit formulation was therefore made (Bunt, 1974, *b*), with the following analysis:

	B	Cu	Fe	Mn	Zn	Mo
Frit WM 255	1·0	4·3	13·8	5·4	4·0	1·0%

8.8 UNITED STATES OF AMERICA

Two groups of composts are widely known and used in the USA.

THE UNIVERSITY OF CALIFORNIA SYSTEM

Professor Baker and his colleagues at the University of California have introduced the UC System of growing. In this system, composts can be made entirely of sand, or peat and sand, or of peat only. There are five different physical mixes and to each of these there are six different base fertilisers, making a total of thirty different composts. In practice only a few of these are used and the following are among the most useful:

UC Mix D

    75% by volume peat moss
    25% by volume fine sand

To this mixture is added either of the following fertilisers:

Fertiliser I(D): this contains only a low amount of inorganic nitrogen.

per cubic metre		per cubic yard
0·15 kg	Potassium nitrate	4 oz.
0·15 kg	Potassium sulphate	4 oz.
1·2 kg	Superphosphate	2 lb.
3·0 kg	Dolomite limestone	5 lb.
2·4 kg	Calcium carbonate	4 lb.

This gives an added nutrient concentration of:

N	P	K
20	95	123 mg/l

An alternative base fertiliser mixture which contains some nitrogen in a reserve form as well as some inorganic nitrogen is:

Fertiliser II(D)

per cubic metre		per cubic yard
1·5 kg	Hoof and horn	2 lb. 8 oz.
0·15 kg	Potassium nitrate	4 oz.
0·15 kg	Potassium sulphate	4 oz.
1·2 kg	Superphosphate	2 lb.
3·0 kg	Dolomite limestone	5 lb.
1·2 kg	Calcium carbonate	2 lb.

These rates of fertilisers supply:

	N	P	K
(mineral)	20	95	123 mg/l
(organic)	195		

UC *Mix E*. This is an all-peat compost and, again, fertilisers can be used at two rates.

Fertiliser I(E): this contains inorganic nitrogen only.

per cubic metre		per cubic yard
0·2 kg	Potassium nitrate	6 oz.
0·6 kg	Superphosphate	1 lb.
1·5 kg	Dolomite limestone	2 lb. 8 oz.
3·0 kg	Calcium carbonate	5 lb.

The added nutrient concentration is:

N	P	K
26	48	76 mg/l

Fertiliser II(E): this contains some reserve or organic nitrogen as well as inorganic nitrogen.

per cubic metre		per cubic yard
1·5 kg	Hoof and Horn	2 lb. 8 oz.
0·2 kg	Potassium nitrate	6 oz.
0·6 kg	Superphosphate	1 lb.
1·5 kg	Dolomite limestone	2 lb. 8 oz.
3·0 kg	Calcium carbonate	5 lb.

This supplies:

	N	P	K
(mineral)	26	48	76 mg/l
(organic)	195		

No micro-element additions are recommended for these composts. An interesting feature of these formulations is the reduction in the amount of liming materials as the sand is decreased and the peat increased. It appears that the fine sand in California is often slightly acidic in reaction; the rate of superphosphate is also reduced as the sand content of the compost is decreased. Steam sterilisation of the compost is generally recommended.

CORNELL PEAT–LITE MIXES

The second group of widely known loamless composts in the USA is the Cornell Peat–Lite Mixes. These mixes were introduced by Professors Boodley and Sheldrake of Cornell University, New York, and are based on mixtures of peat and vermiculite or peat and perlite. The fertiliser levels are varied slightly, depending upon the type of plant to be grown and whether liquid feeding or slow-release fertilisers are to be used.

Peat–Lite Mix A for pot plants comprises:

50% by volume sphagnum peat
50% by volume vermiculite.

To this is added:

per cubic metre		*per cubic yard*
0·9 kg	Calcium or Potassium nitrate	1 lb. 8 oz.
0·6 kg	Superphosphate	1 lb.
3·0 kg	Ground limestone	5 lb.
0·07 kg	Fritted trace elements (FTE No. 503)	2 oz.

This supplies:

N	P	K
117	48	340 mg/l

Peat–Lite Mix B, the vermiculite is replaced with perlite.

50% by volume sphagnum peat
50% by volume perlite

The fertilisers are the same as for Mix A, except that potassium nitrate must be used and not calcium nitrate; this is because the vermiculite contains available potassium whereas the perlite does not. The amount of superphosphate added must also be doubled.

The fritted trace element mixture has an analysis of:

B	Cu	Fe	Mn	Zn	Mo
3	3	18	7·5	17	0·2%

With these two mixes, no slow-release fertilisers are added and liquid feeding would start 2–3 weeks after potting. If a complete NPK slow release fertiliser is required, 'Osmocote' 14–14–14 is used at the rate of 5 lb./cu.yd (3 kg/cu.m).

A feature of the Cornell Peat–Lite mixes is the use of a non-ionic wetting agent, three fluid ounces being added to 5–10 galls of water for each cubic yard of mix (85 ml to 30–60 l/cu.m). Alternatively, the wetting agent can be added first to a small amount of vermiculite, which is then added to the bulk mix.

Recently, two additional mixes have been introduced for foliage and epiphytic plants.

Cornell Foliage Plant Mix is recommended for plants which require a compost with high moisture retention characteristics. The bulk mix comprises:

2 parts by volume sphagnum peat
1 part by volume vermiculite (grade No 2)
1 part by volume perlite (medium grade).

To this is added:

per cubic metre		*per cubic yard*
0·6 kg	Potassium nitrate	1 lb.
1·2 kg	Superphosphate	2 lb.
1·6 kg	10–10–10 fertiliser	2 lb. 12 oz.
4·9 kg	Dolomite limestone	8 lb. 4 oz.
0·4 kg	Iron sulphate	12 oz.
0·07 kg	Fritted trace elements	2 oz.

This supplies the following nutrients:

N	P	K
238	165	360 mg/l

This mix is recommended for such plants as: *Aphelandra squarrosa, Begonia, Beloperone guttata, Cissus, Coleus,* Ferns, *Ficus, Hedera, Maranta, Pelargonium, Pilea, Sansevieria.*

Cornell Epiphyte Mix is used for plants that require good drainage and aeration and are able to withstand drying out between waterings. The bulk mix comprises:

1/3 by volume sphagnum peat (screened $\frac{1}{2}$ in. mesh)
1/3 by volume Douglas red or white fir bark ($\frac{1}{8}$ in.–$\frac{1}{4}$ in.)
1/3 by volume Perlite (medium grade)

To this is added:

per cubic metre		*per cubic yard*
0·6 kg	Potassium nitrate	1 lb.
2·7 kg	Superphosphate	4 lb. 8 oz.
1·5 kg	10–10–10 fertiliser	2 lb. 8 oz.
4·2 kg	Dolomite limestone	7 lb.
0·3 kg	Iron sulphate	8 oz.
0·07 kg	Fritted trace elements	2 oz.

This gives an added nutrient concentration of:

N	P	K
228	280	352 mg/l

Plants grown in this mix are: *Bromeliad, Cacti, Crassula, Diffenbachia, Episcia, Gloxinia, Hoya, Monstera, Philodendron, Peperomia.*

If a compost with slow-release nutrients is required, the fertilisers other than Dolomite limestone and superphosphate are omitted and either 'Osmocote' 14–14–14 or Peter's 14–7–7 added at 5 lb./cu.yd (3 kg./cu.m).

PENNSYLVANIA STATE PEAT FORMULAE

Recently two composts for propagating and growing ornamentals have been introduced by the Pennsylvania State University (White, 1974). They are both based on sphagnum peat.

The Propagation Mix contains:

per cubic metre		per cubic yard
0·9 kg	Superphosphate	1 lb. 8 oz.
0·3 kg	Potassium nitrate	8 oz.
0·15 kg	Complete fertiliser (20–19–18)	4 oz.
3·0 kg	Dolomite limestone	5 lb.

This gives a nutrient concentration of:

N	P	K
69	83	136 mg/l

The Growing-on Mix contains:

per cubic metre		per cubic yard
1·2 kg	Superphosphate	2 lb.
0·6 kg	Potassium nitrate	1 lb.
0·6 kg	Complete fertiliser (20–19–18)	1 lb.
3·0 kg	Dolomite limestone	5 lb.

The nutrient concentrations are:

N	P	K
192	143	317 mg/l

The $20N-19P_2O_5-18K_2O$ fertiliser is water-soluble and can be used also to make up liquid feeds. It contains micro-elements, mainly in chelated form.

8.9 SAWDUST AND BARK COMPOSTS

Increasing attention is being given by growers to the use of sawdust and pulverised bark in potting composts. These materials are bulky and not as easily transported as compressed and baled peat, and for that reason interest is usually centred around areas of forestry production or wood processing. Whilst there is not as much experimental data on the use of these materials as there is on peat-based composts, the principles governing their satisfactory use for compost making are generally understood and have been discussed in Chapter 2. The principal requirement is to adjust the amount of nitrogenous fertiliser used in relation to the type of wood and the treatment it has received prior to its being used in the compost. Further work is, however, required on the subsequent liquid feeding programmes necessary for growing a wide range of ornamental plants in these materials.

SOFTWOOD SAWDUST COMPOSTS

The following three composts are based on recommendations given by Johnson (1968) for the use of redwood sawdust:

Compost I

      66·6% by volume Nitrolised* redwood sawdust
      33·3% by volume fine sand.

To this is added:

per cubic metre		per cubic yard
1·48 kg	Superphosphate	2 lb. 8 oz.
0·30 kg	Potassium sulphate	8 oz.
1·78 kg	Dolomite lime	3 lb.
0·59 kg	Iron sulphate	1 lb.

Compost II

      33·3% by volume raw redwood sawdust
      33·3% by volume sphagnum peat
      33·3% by volume fine sand

To this is added:

per cubic metre		per cubic yard
0·59 kg	Calcium nitrate	1 lb.
1·48 kg	Superphosphate	2 lb. 8 oz.
0·59 kg	Potassium nitrate	1 lb.
4·15 kg	Calcium carbonate	7 lb.
4·15 kg	Dolomite limestone	7 lb.
0·59 kg	Iron sulphate	1 lb.

This mixture must be stored for a minimum of two weeks before it is used for potting.

Compost III

      66·6% by volume raw redwood sawdust
      33·3% by volume fine sand

---

*Nitrolised is a copyright name meaning 'nitrogen stabilised'. The product consists of redwood sawdust which has been previously composted with 0·5% nitrogen, based on the dry weight of the sawdust.

164 Compost Formulation and Preparation

To this is added:

per cubic metre		per cubic yard
1·19 kg	Calcium nitrate	2 lb.
1·48 kg	Superphosphate	2 lb. 8 oz.
0·59 kg	Potassium nitrate	1 lb.
2·97 kg	Calcium carbonate	5 lb.
4·15 kg	Dolomite limestone	7 lb.
0·59 kg	Iron sulphate	1 lb.

This mixture must also be stored for at least two weeks before it is used for potting.

SHREDDED OR PULVERISED HARDWOOD BARK COMPOSTS
Gartner *et al.* (1973) make the following recommendation for a compost made from a mixture of hardwood barks and sand.

The bark is first ground to give the following particle size distribution:

Size	% by weight
$> \frac{1}{4}$ in. (> 6.4 mm)	0·7
$\frac{1}{4}-\frac{1}{8}$ in. (6·4–3·2 mm)	12·9
$\frac{1}{8}-\frac{1}{16}$ in. (3·2–1·6 mm)	32·6
$\frac{1}{16}-\frac{1}{32}$ in. (1·6–0·7 mm)	17·6
$\frac{1}{32}-\frac{1}{50}$ in. (0·7–0·5 mm)	9·7
$< \frac{1}{50}$ in. (< 0.5 mm)	26·5

The bulk mix is then prepared from:

66·6% by volume ground bark
33·3% by volume coarse sand.

To this is added:

per cubic metre		per cubic yard
3·56 kg	Ammonium nitrate	6 lb.
2·97 kg	Superphosphate	5 lb.
0·59 kg	Elemental sulphur	1 lb.
0·59 kg	Iron sulphate	1 lb.

This is mixed thoroughly in a rotary mixer and then kept in a stockpile for 30 days, during which time it must be kept moist (60–90% moisture). No further fertiliser is added before the mixture is used, but liquid feeding of a 20–20–20 fertiliser at 250 N, 110 P, 207 K (ppm) is given at each irrigation. Analysis of the bark for micro-elements gave:

aluminium 1143, boron 24, iron 743, manganese 485, sodium 548, and zinc 53 (ppm) and no further micro-element additions are recommended.

Cappaert *et al.* (1974) experimented with a mixture of hardwood barks for growing ornamental plants. The bark was first composted for six weeks with the following nutrient additions (by weight):

0·6% nitrogen (as ammonium sulphate), 0·3% P, 0·2% K.

Four weeks after the start of composting, practically all of the mineral nitrogen that had been added had been converted to organic nitrogen, and at the finish of composting, 50% of the bark particles were less than 1 mm in diameter. Experimental results showed that the organic nitrogen present in the bark after composting was not readily released to the plant and further nitrogen was necessary. It was recommended that either 1 kg/cu.m (1·68 lb./cu.yd) of ammonium nitrate be mixed into the compost before potting or 2 kg/cu.m (3·37 lb./cu.yd) of a 20–10–15–6 slow release fertiliser be used in order to obtain the satisfactory rates of plant growth.

SOFTWOOD BARK COMPOSTS

Softwood bark is much more resistant to bacterial decomposition than hardwood bark. Even after the softwood bark has been stockpiled or composted for twelve weeks to allow the biological degradation of any toxic substances it may contain, the particles will not have been broken down significantly; neither will there be a rapid increase in the rate of breakdown when nutrient solutions are applied to the bark, as is the case with some sawdusts and hardwood barks. This resistance to decomposition, together with the natural resistance of the particles to water absorption, means that a container filled with only shredded softwood bark will drain quickly after irrigation and will not retain as much available water as will some other composts.

A typical sample of the pulverised bark available to growers in Britain had the following particle size analysis after it had been stockpiled for 12 weeks:

>5 mm	24·19% (of dry weight)
5–3 mm	18·13%
3–2 mm	16·30%
2–1 mm	15·00%
<1 mm	26·3%

The standard test for air and water retention showed that this material had an available water content of approximately 58% and an air capacity of 30%. When a mixture of 50% by volume pulverised bark and 50% of sphagnum peat was made, the available water content had increased to 64% and the air capacity had decreased to 22%. Comparable values for a mixture of 75% sphagnum peat and 25% fine sand are approximately 70% for the available water content and 10% for the air capacity.

For these reasons, composts made exclusively from softwood bark are used only for epiphytic plants or in certain cases where very good drainage

166 *Compost Formulation and Preparation*

is required. Usually, the pulverised bark is mixed in varying ratios with sphagnum peat to increase the water retention of the mixture. Used in this way, the fertiliser additions are similar to those of other peat-based composts.

8.10 AZALEA COMPOSTS

Azaleas will grow satisfactorily only in a compost which has a relatively low pH, i.e. in the range 4·5 to 5·5; also the salinity or nutrient levels must be lower than is customary for most plants. For these reasons, composts which are used for the majority of pot plants are not acceptable for azalea growing. Nutritional studies made with plants grown in sand cultures have shown that the azalea grows best when most of the nitrogen is in the ammonium form. This form of nitrogen reduces the uptake of other cations and helps to maintain a low pH within the leaf tissue. The availability of the iron is thereby increased and the incidence of iron deficiency, a disorder to which azaleas are particularly susceptible, is reduced. The pH of the compost must also be kept low during the growing season. In areas having water with a high calcium content, some of the carbonates must be neutralised by treating the water with phosphoric acid; alternatively, acidifying liquid feeds, based largely on ammonium sulphate, can be used to prevent the rise in the compost pH which would otherwise occur. It must also be remembered that, at the low pH values maintained in azalea composts, boron toxicity can develop quickly if too much boron is present in either the water or the fertilisers used to make up the liquid feed. Azaleas are also very susceptible to salt injury arising from high levels of fertilisers. The injury is seen first as a yellowing of the leaves which later turn brown and drop off. Composts having a poor physical structure with a low air capacity, and which remain excessively wet after being irrigated, can also cause the leaves to drop.

In Belgium and the Netherlands, the traditional material for azalea composts has been pine needles. The rapid expansion in the number of azaleas grown, however, has forced growers into using other media, principally sphagnum peat, sometimes with a quantity of shredded bark or foam plastic added to improve the aeration. These media have lower levels of available nutrients than pine needles compost and more attention must be given to the liquid feeding.

BELGIUM

Blomme and Piens (1969) analysed the composts in which a large number of azaleas of different ages were growing, and related the analysis to the growth of the plants. They concluded that the most favourable levels of nutrients in the compost were: 11 mg of phosphorus, 33 mg of potassium and 357 mg of calcium per 100 g of compost. Magnesium levels did not appear to influence growth and no data were given for nitrogen levels.

Also in Belgium, Gabriels, Engles and van Onsem (1972) used the systematic variation technique of Homes (1963) to determine the optional balance of anions and cations for azalea cv. 'Ambrosiana' grown in peat or in coniferous litter. Peat grown plants were considered superior to those grown in the coniferous litter and the best results were obtained when a total of 2·08 g N, 0·35 g P, 2·72 g K, 1·61 g Ca, 0·98 g Mg and 0·15 g S was applied to each plant over the growing season, by giving liquid feeds at weekly intervals. From these data, the optimal ratio of the three main nutrients in the fertiliser can be calculated as 1 N; 0·17 P; 1·3 K.

The analysis of the compost at the end of the growing season was:

pH 5·1 (in water), 4·5 (in KCl), conductivity, 0·427 millimhos.

The extractable nutrients were:

NO_3 201, P 38, K 190, Ca 655 and Mg 270 mg/l.

Mineral composition of the leaf tissue was:

N 2·6%, P 0·49%, K 2·0%, Ca 0·75%, Mg 0·28%, SO_4 0·71%.

NETHERLANDS

Arnold Bik (1972) studied the nutrition of azalea cv. 'Ambrosius' during the period when the young plants are grown in beds out of doors, prior to being potted up and brought into the glasshouse for forcing, i.e. from the end of May to the end of September. The plants were grown in beds of decomposed sphagnum peat to which magnesium and micro-elements were added prior to planting. All other nutrients were applied in liquid feeds on three occasions during the season. It was concluded that the best results were obtained when the total amounts of nutrients applied during the season were:

N 72, P 7·8, K 19·9 g/sq.m of bed.

This is equivalent to a total of:

194 g ammonium nitrate
37·4 g monoammonium phosphate
48·5 g potassium nitrate

per square metre supplied in the three liquid feeds.

The leaves of the plants receiving these rates of nutrients had the following mineral analysis:

N 2·19%, P 0·18%, K 0·70%.

The organic salt content of the leaves, defined as the sum of the cations K + Na + Ca + Mg, minus the sum of the anions NO_3 + H_2PO_4 + SO_4 + Cl, was found to be only one half of the value in chrysanthemum leaves and only one third of the value in gloxinia leaves, thereby showing a distinct ionic behaviour for this species.

UNITED STATES OF AMERICA

In the USA Batson (1972) has recommended a coarse grade of sphagnum peat with a pH of 5·0 as the growing medium. No base fertiliser is mixed with the peat, apart from chalk if it is necessary. Feeding is commenced two weeks after potting and then repeated at weekly intervals, one of three formulations of liquid feed being used. The choice depends upon the pH of the compost, which is determined at weekly intervals.

If the compost pH is between 4·9 and 5·4, the following liquid feed is used; Feed 'A' (23–10–12):

g/100 litres		oz./100 US galls
78·6	ammonium nitrate	10·5
45	diammonium phosphate	6
45	potassium nitrate	6

This solution has approximately:

N 430, P 100, K 170 ppm.

If the compost pH rises above 5·5, use Feed 'B' (21–0–0):

g/100 litres		oz./100 US galls
180	ammonium sulphate	24
120	iron sulphate	16

This feed is acidifying and will reduce the pH; it has N 378 ppm.

If the compost pH drops below 4·8, the following feed is used to raise the pH again, Feed 'C' (15–3–3):

g/100 litres		oz./100 US galls
314	calcium nitrate	42
22·5	monocalcium phosphate	3
30	potassium nitrate	4

This feed contains N 511, P 54, K 114 ppm.

No feed must ever be given when the compost is dry; an irrigation with plain water should be given first. Feeding is discontinued for seven days preceding each occasion when the plants are pinched or stopped and is not resumed until the new growth is 0·5 in. (1·25 cm) long. The salinity of the compost is monitored weekly and, during the period of pinching, it should not exceed a value of 0·6 millimhos on a Solu Bridge test based on 1 part of compost to 5 parts of water; for a test based on one part of compost to two parts of water, the equivalent value would be 1·1 millimhos. During the remainder of the growing period, the salinity values should not exceed 0·90 on a 1 to 5 test or 1·9 on a 1 to 2 test. Regular feeding with iron in chelated form is given at 4 oz./100 galls (30 g/100 l) every month and always with the first feed after every pinch. Nutrient levels should be determined each month and on the spurway system of analysis should be about 25–30 N, 3–5 P, 15–20 K and 100–150 Ca ppm.

8.11 PROPRIETARY FORMULATIONS

Several proprietary loamless composts are on sale in the United Kingdom but, as their formulation is usually not disclosed, little can be said about them. The convenience of using ready mixed or prepared composts has to be balanced against their higher costs, in some cases they are much more expensive than the cost of home mixing. Much will depend upon the quantity of compost required, the amount of machinery available for mixing and also the labour situation. Frequently, the source of nitrogen used in proprietary composts is not disclosed. One potential danger in using composts with an unknown composition which have been stored for an unknown period lies in the possible breakdown of organic or slow-release types of fertiliser into ammonium nitrogen, during the period between the time of mixing the compost and its use. Problems arising from ammonium nutrition under the low light conditions of winter have been discussed in Chapter 5. In winter, a large part of the nitrogen should be in the nitrate form.

8.12 COMPOST PREPARATION

Because of the different amounts of compost required and the variation in the amount of equipment and facilities available on each nursery, no one standardised procedure for compost preparation can be given. Two objectives should, however, be common to all systems:

(1) uniformity and consistency in the product,
(2) maximum mechanisation and low labour input.

Uniformity and consistency can be achieved by selection and quality control of the basic ingredients and in adequate mixing. Uneven levels of nutrients can be caused by either inefficient mixing or by powdered fertilisers such as superphosphate, potassium nitrate, ammonium nitrate, etc., being added to peat which is too wet. When this occurs, the fertilisers tend to adhere to the first particles of wet peat they touch and even prolonged mixing will not then give an even nutrient distribution. The moisture content of sphagnum peat should not exceed 250% of the oven-dry weight of the peat (approximately 70% of the weight of the moist sample of peat) if good uniform mixing is to be achieved.

STERILISING EQUIPMENT

Materials used to make loamless composts do not usually require partial sterilisation. If, however, this is considered necessary or if a loam-based compost is being prepared, heat sterilisation is preferable to chemical treatment. The best method of steaming soil is to place it over a pipework grid or perforated metal plate and allow the steam and the displaced air to rise up through the soil as steaming progresses. This is a more efficient method than injecting the steam under a plastic sheet and using pressure to force the steam down into the soil.

The container or bin in which the steaming is done can be mobile or static; usually less labour is required to operate mobile bins. For efficient sterilisation, a steam separator must be fitted at the end of the steam main; this removes excessive amounts of any free water present in the steam and so prevents 'puddling' of the soil and uneven heating.

Working with steam–air mixtures instead of steam, workers at Pennsylvania State University (Aldrich et al. 1972) have found that placing the compost in a box with a perforated base and then injecting the steam–air mixture from the top, either under a plastic sheet which is tightly tied down or under an airtight lid, gives better results then applying the steam–air mixture to the bottom surface and then allowing it to penetrate up through the compost. If there is any unevenness in the soil structure or if the pressure of the mixture is too great, injection of the steam–air mixture from the base can result in a 'chimney' effect, leaving areas of compost untreated. This does not occur when the mixture is injected from the surface. It is, however, necessary to ensure that adequate exits for the mixture are provided at the base of the container, e.g. a gravel base, a perforated plate or spaced boards at the bottom of the bench.

The air required to make the steam–air mixture is best supplied by a small mobile power driven fan or aerator, rather than relying on a high pressure steam injector to entrain sufficient free air to give the desired temperature in the mixture.

MIXING

Compost preparation can be done either in batches or by a continuous process. The former system offers the greatest accuracy and uniformity of fertiliser mixing and is recommended when composts are prepared on the nursery. The continuous process system offers the greatest productivity but can be satisfactorily employed only under closely supervised industrial conditions.

There are several types of mixers that can be used on the nursery for compost mixing, examples of three types are shown in Figures 8.1, 8.2 and 8.3. The mixer in Figure 8.1 has a stationary horizontal drum with rotating blades. First, a bale of compressed peat is fed into the mixer to be loosened and broken up; then a measured quantity of sand is added plus the pre-weighed fertilisers. After two minutes of mixing in the dry state, the compost can be brought to the desired water content by adding water through a built-in

Compost Preparation 171

spray line, whilst the mixer is still working. The compost is then discharged from the bottom of the drum into either polythene sacks or wheeled trucks for conveyance to the potting area. In the system illustrated, the compost is discharged from the mixer on to an endless belt which conveys it to the hopper of an automatic pot-filling machine; this system allows 2 000 14-cm pots to be filled per hour. With this type of mixer, a cubic yard of compost can be prepared in about 10 minutes, whereas hand mixing requires about 1·75 man-hours.

Fig. 8.1 Mechanised mixing and pot filling operation. The 'Adelphi' mixer (on the left) has a stationary horizontal drum with rotating mixing blades. The mixed compost is discharged on to the endless belt elevator, which conveys the compost to the 'Javo' potting machine. This system enables 2 000 14 cm pots to be filled per hour. (Acknowledgement to Framptons Ltd, Chichester, England.)

Rotary or revolving drum type of concrete mixers are available in a range of capacities. Large machines, as illustrated in Figure 8.2, are fitted with mechanical loaders. A third type of mixer, illustrated in Figure 8.3, employs a different principle. The materials in the hopper are mixed by being agitated with a slatted conveyor belt working on an inclined base. The two latter types of mixer can be mobile if desired. Shredding machines do an excellent job of shredding but they do not perform efficiently as mixers, and neither can good mixing be obtained by running a rotary cultivator over a heap of materials spread on to the concrete floor of a shed. Mixers that tend to grind or pulverise the compost and so destroy the structure should also be avoided. At present

172 *Compost Formulation and Preparation*

Fig. 8.2 Concrete-type rotary mixer. This machine is filled by placing the materials in the 'skip' (on the left) which is raised by a hydraulic ram to discharge into the drum. After mixing and adding water, the compost is discharged into a trailer (on the right) by reversing the direction of rotation. (Acknowledgement to Southdown Flowers Ltd, Yapton, England.)

Fig. 8.3 Alvan Blanch meal mixer used for compost mixing. Mixing is by means of a slatted conveyer which moves the materials up the inclined base towards a segmented agitator which gives a lateral movement to the ingredients. This continuous circulatory movement mixes the compost in approximately five minutes. Water can be added after mixing, by a spray-bar coupled to the water supply. The mixer illustrated is fitted with a potting attachment. (Acknowledgement to Alvan Blanch Ltd, Malmesbury, Wilts, England.)

prices, it has been estimated that the cost of an efficient mixer will be covered by the saving in labour charges, if 100 cu.yd of compost are mixed per annum.

POTTING MACHINES

These range from the small mobile, single-operator 'Potmore' types of bench, where pot filling and potting is done manually, to the large machines, where filling and potting is automatic, e.g. 'Javo', 'Plantarex', etc. Whilst a team of 3–4 workers is required to operate these machines, it has been estimated (Stoffert, 1973) that, at the costs prevailing in Germany in 1968, repotting by machine was cheaper than repotting by hand when the number of plants to be repotted exceeded 150 000 per annum. It has also been estimated that the large anticipated increase in labour costs means that, by 1980, the number of plants required to make machine potting cheaper than hand potting will have fallen to 100 000 per annum.

Chapter 9

Liquid feeding

When plants are grown for long periods in small volumes of compost, the supply of available nutrients becomes depleted, growth ceases and symptoms of nutrient deficiency may spoil the general appearance of the plants. Years ago, the grower's remedy was either to occasionally add to each pot a teaspoonful of powdered fertiliser which was washed into the compost at the next watering, or to make up a liquid feed by suspending animal manure in coarse sacking in a barrel of water and allowing the soluble nutrients to diffuse into the water. This solution was then applied to the pots with a watering can. These methods of liquid feeding are too imprecise and time consuming to be used today and the practice of adding soluble chemicals to the irrigation water has steadily gained in popularity since it was first used in the 1930s. Factors contributing to the growth of this system of liquid feeding have been:

(1) the large increase in the cost of labour;
(2) the need to obtain the maximum growth rate and the greatest number of plants per year from each glasshouse;
(3) the development of automatic or semi-automatic systems of pot watering which, together with fertiliser injectors or diluters, provide an efficient and easy means of combining feeding with watering;
(4) the increased use of loamless composts, which have a greater need for liquid feeding than the older, loam-based composts.

9.1 IMPORTANCE OF LIQUID FEEDING

How important is liquid feeding when using loamless composts? This is a deceptively simple question which cannot be answered without first considering the following points:

(a) The type of plant, its rate of growth and size in relation to the size of the container. The chrysanthemum is an example of a quick growing plant having a large nutrient requirement; Saintpaulia is an example of a plant with a much lower nutrient requirement.

(b) The type of fertiliser used in preparing the compost. If slow-release types of fertiliser are used, larger amounts of nutrients can be added in the base fertiliser and this reduces the importance of liquid feeding.
(c) The system of irrigation used. Pots receiving any form of surface watering are likely to lose some nutrients by leaching, whereas with the capillary system of irrigation, there will be no loss of nutrients by leaching.

Obviously no one answer will apply to all situations. Some growers rely entirely upon slow-release fertilisers to supply all the nutrients required by the plants and do not use liquid feeds at all. Crops suitable for this system of growing are pilea and calceolaria, which do not make much growth (dry weight) in relation to the size of the pot. Vigorously growing plants, e.g. aphelandra, or plants which are slow to reach maturity, e.g. cyclamen, will give better results with liquid feeding, even when some slow-release fertilisers have been used in preparing the compost.

BASE FERTILISERS VERSUS LIQUID FEEDING

The importance of liquid feeding in pot chrysanthemum culture is shown from some experimental results obtained at the Glasshouse Crops Research Institute (Table 9.1). Plants were grown in a peat–sand mixture in the normal commercial way with five rooted cuttings per 14 cm ($5\frac{1}{2}$ in.) pot, each holding one litre of compost. In one treatment, the nitrogen, phosphorus and potassium were supplied only from the base fertiliser; the plants did not receive any liquid feeding. In a second treatment, the plants were given N, P and K only in the liquid feed and none in the base fertiliser, whilst in a third treatment the plants had N, P and K from both the base fertiliser and the liquid feed. There was also a fourth treatment, in which the plants were not given any N, P and K in either the base fertiliser or the liquid feed. With all treatments calcium carbonate, Dolomite limestone and a fritted trace element mixture were added as a base fertiliser dressing.

Comparison of the amount of dry weight produced by the different treatments, Table 9.1, shows that those plants which had N, P and K in only the liquid feed made considerably more growth than plants which had N, P and K in only the base fertiliser and no liquid feeding. In terms of the total amounts of nutrients absorbed by the plants, the liquid feed treatment had an even greater advantage over the base fertiliser treatment. In summer, plants having the liquid feed and no base fertiliser contained approximately $2\frac{1}{2}$ times the amount of N and K, and $1\frac{1}{2}$ times the amount of P, of plants which had been given only the base fertiliser and no liquid feeding. Although the plants grown with the base fertiliser and no liquid feeding had achieved a reasonable amount of growth, i.e. dry weight, they had considerably lower concentrations of N and K in their tissue than the liquid-fed plants. They

Liquid Feeding

Table 9.1. Effect of base fertiliser and liquid feeding on the growth and composition of pot chrysanthemums.

Fertiliser Treatment	Summer					Winter				
	Dry weight (g)	Total uptake (mg)			leaf N %	Dry weight (g)	Total uptake (mg)			leaf N %
		N	P	K			N	P	K	
No base or liquid feed	5.4	82	4	71	2.0	4.7	129	7	118	2.6
Base only	36.4	608	79	583	2.3	14.3	502	62	479	3.9
Base and liquid feed	56.9	1870	171	1833	4.8	20.3	1093	119	1160	5.9
Liquid feed only	46.4	1494	127	1485	4.8	23.4	1118	106	1263	5.7

looked underfed or starved, which adversely affected their general appearance and quality more than the dry weight figures would suggest. In summer, the leaves of these plants were pale green with only 2·3% nitrogen, whilst the liquid-fed plants had large, deep-green leaves with 4·8% nitrogen. Under the lower light levels of winter, all treatments resulted in less growth, but the results were essentially the same as for the summer experiment.

These results illustrate the great importance of liquid feeding in the culture of such crops as chrysanthemum, which have a high requirement for nutrients by comparison with other slower growing plants (Table 5.1). The extent of the benefit from liquid feeding will, of course, depend very much upon the relative amounts of nutrient being supplied by the base fertiliser and the liquid feed, and also on the amount of nutrients lost from the compost by leaching. In the experiments quoted above, the plants were given the normal amounts of base fertilisers (p. 156) apart from the nitrogen which was supplied as hoof and horn. The liquid feed was also of normal strength with 200 N, 15 P, 150 K and 15 Mg ppm applied at each irrigation, commencing at two weeks from potting and maintained until the buds showed colour.

INTERACTION OF SEASON AND STRENGTH OF FEED

Crops grown with automated water and feeding systems normally receive fertiliser solution at each irrigation. Often the plants are never given any plain water at all and in these circumstances it is important to use the correct strength of feed. If it is too strong, salinity problems can occur, even though there is some loss of nutrients from the pot by leaching. Typical results showing the effects of the levels of nitrogen and potassium in the liquid feed on the salinity of the compost are shown in Figure 9.1. The phosphorus level was kept at 15 ppm for all feeds, which were applied at each irrigation commencing 14 days after potting. In winter, the weakest feed of 50 N and 50 K (ppm) failed to maintain the initial salinity value of the compost at EC_e 3 millimhos, the final value being EC_e 1·5 millimhos. With the 200 N, 200 K feed, the salinity value of the compost remained approximately constant over the life of the crop, whilst at 400 N 400 K it rose to 4·5 millimhos during the 10 week growing period. In summer, these same liquid feeds had a much greater effect upon the salinity of the compost. The final salinity value of the 50 N 50 K treatment was only slightly less than the initial value, whilst, with the other feeds, there was a progressive increase in the compost salinity during the experiment. The final value for the 200 N 200 K treatment was almost double the initial value, whilst the 400 N 400 K feed gave a final value of EC_e 7·9 millimhos, which was almost three times the starting value and equivalent to an osmotic pressure of almost 2·5 atmospheres. As the water content of the compost in the pots could have fallen between irrigations to only half of the amount present when making the salinity determinations, the salinity of the compost would

178 Liquid Feeding

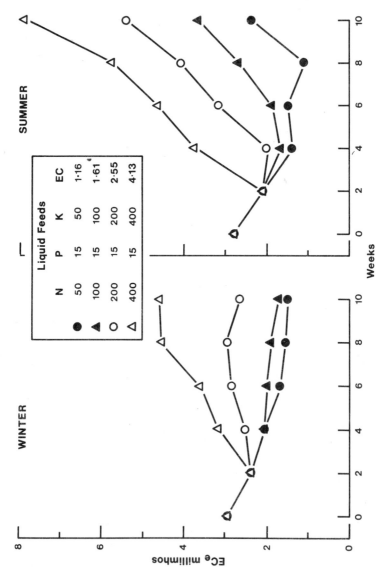

Fig. 9.1 Effect of the strength of the liquid feed and the season on the compost salinity of pot-grown chrysanthemums. Note the higher compost salinity in summer than in winter and the rapid increase in salinity at the end of the crop.

at times have been equivalent to 5 atmospheres. At this high value there is considerable risk of root damage.

SEASON AND EVAPOTRANSPIRATION RATES

In the above experiment, it was not possible to measure the amount of liquid being applied to the pots or the amount of leaching which occurred. In other experiments, however, the precise amounts of water used in evapotranspiration, i.e. evaporation plus transpiration, have been measured daily. Data from these experiments offer an explanation for the differences found in the salinity values of the summer and winter crops (Fig. 9.1). In mid-winter, the average water requirement of chrysanthemum plants grown in the normal 14 cm plastic pots was found to increase from 16 ml per day during the first week after potting, to 100 ml per day in the last week. The total amount of water and liquid feed used in evapotranspiration over the growing period was 5 105 ml per pot and the total dry weight of plant tissue was 16·15 g. In summer, the average daily water requirement in the first week of growth was 88 ml per pot. This rose progressively to an average value of 375 ml per day during the last week before marketing. The total amount of water lost by evapotranspiration was 16 744 ml and the total plant dry weight 45·3 g per pot. Thus, *when there is no loss of water by leaching,* pot chrysanthemums grown in summer lost more than three times as much water in evapotranspiration as plants grown in winter. Under commercial conditions, the total amount of water and fertilisers applied in summer will therefore need to be three to four times as great as in winter, and this accounts for the higher salinity (Fig. 9.1) and pH (Table 9.6) of summer grown crops. The rapid increase in the amount of water and fertilisers applied as the plant approaches maturity is also the reason for the build-up in salinity at the end of the summer crop. Values given for the average daily water requirement are largely determined by the amount of solar radiation received. On dull days, the water requirement may be only half the average value, whilst on bright days it can be double the average value; evapotranspiration losses as great as 550 ml per pot per day have been recorded on bright days in summer. In terms of the irrigation of border or field crops, this is equivalent to a water requirement of 3·5 cm (1·4 in.) of water per day.

In the capillary or sub-irrigation system of watering (p. 208), the pots are not leached. If liquid fertilisers are added to the irrigation water, there will be the risk of an increase in the compost salinity and nutrient levels, unless the strength and frequency of the feeding is related to the needs of the plants. Bunt (1973, *b*) found that the flowering of winter-grown pot chrysanthemums was delayed by the build-up of the mineral nitrogen in the compost. A liquid feed containing 300 ppm N applied to the surface of pots having free drainage had the same effect, in terms of plant response and nitrogen levels in the compost, as a feed of only 100 ppm given to plants grown on a capillary bench.

9.2 FORMULATING LIQUID FEEDS

Pot plant growers have the choice of using either proprietary liquid feeds or making up the feeds themselves from soluble chemicals. It will be assumed that in either case the feed will be in a concentrated form and will then be diluted by means of suitable equipment.

PROPRIETARY FEEDS

One of the principal advantages of using proprietary feeds is convenience. The concentrated feed is already prepared and it is not necessary to weigh out and dissolve the chemicals, although in some cases the feeds are supplied in powder rather than in liquid form. Advantage can also be taken of the advanced technology used by the manufacturers in formulating the fertilisers. Some of the materials they use are either not available or are not practical for growers to use. For example, phosphoric acid and ammonia frequently form the basis of proprietary liquid feeds, with urea being used as a secondary source of nitrogen. Other materials used by commercial manufacturers are ammonium hydroxide, chelated macro- and micro-elements and growth regulators such as indolbutyric acid. In the USA, superphosphoric acid and ammonium polyphosphates are also used in some proprietary feeds. The advantages of using such feeds are: higher ratios of plant nutrients, resistance to 'salting-out' at low temperatures and the prevention of minor element precipitation by chelation. Some care must be exercised, however, over the use of liquid feeds containing high concentrations of chelated micro-elements; for example, manganese toxicity could result if the feed contains too much chelated manganese. Present experience suggests that the amount of chelated manganese in the diluted feed should not be more than about 0·5 ppm. Feeds containing chelated micro-elements should not be used indiscriminately; they should be used only when there is evidence of a deficiency.

In the USA, it is possible to purchase a wide range of liquid feeds formulated specifically for certain crops or for use with special composts. For example, Peters Fertilisers of Pennsylvania, USA, produce, for azalea growers, a liquid feed with a strongly acidifying action. A 'Peat–Lite Special' fertiliser with 20N, $19P_2O_5$, $18K_2O$ (20 N, 8·4 P, 15 K) is also formulated for use with peat-vermiculite and peat–perlite composts.

NURSERY MIXED FEEDS

An alternative to buying proprietary, ready-mixed feeds is to make up concentrated fertiliser solutions in the nursery. These have the advantage that (a) they are cheap and (b) it is then easier to follow non-commercial advice obtained from National or State Advisers. Growers sometimes have difficulty translating liquid feed programmes given by advisers into terms of commercial liquid feeds. By dissolving the correct amounts of a few chemicals in water, it is possible for the grower to prepare, easily and cheaply, liquid feeds that meet most of his requirements. Some of the materials which can be used to make the feeds and their solubilities are listed in Table 9.2. Liquid feeds suitable for most purposes can be formulated from ammonium nitrate, monoammonium phosphate and potassium nitrate. Urea can be used in place

Table 9.2. *Solubility of fertilisers in cold and hot water.*

Fertiliser	Solubility g per 100 cc water		pounds per gallon of cold water
	Cold	Hot	
Ammonium nitrate	$118.3^{(0)}$	$871.0^{(100)}$	11.7
Ammonium sulphate	$70.6^{(0)}$	$103.8^{(100)}$	7.0
Calcium nitrate	$102.5^{(0)}$	$376.0^{(100)}$	10.0
Urea	78.0		7.5
Monoammonium phosphate	$22.7^{(0)}$	$173.2^{(100)}$	2.2
Diammonium phosphate	$57.5^{(0)}$	$106.0^{(70)}$	5.6
Potassium carbonate	$112.0^{(20)}$	$156.0^{(100)}$	11.0
Potassium chloride	$34.7^{(20)}$	$56.7^{(100)}$	3.4
Potassium nitrate	$13.3^{(0)}$	$247.0^{(100)}$	1.3
Potassium sulphate	$12.0^{(25)}$	$24.0^{(100)}$	1.2
Potassium orthophosphate	$90.0^{(20)}$		8.9
Monopotassium phosphate	$167.0^{(20)}$		16.5
Magnesium sulphate	$26.0^{(0)}$	$73.8^{(100)}$	2.5
Sodium borate (Borax)	$1.6^{(10)}$	$14.2^{(55)}$	0.15
Solubor	$4.5^{(10)}$	$32.0^{(50)}$	0.44
Copper sulphate	$31.6^{(0)}$	$203.3^{(100)}$	3.1
Manganese sulphate	$105.3^{(0)}$	$111.2^{(54)}$	10.4
Ferrous sulphate	15.6	$48.6^{(50)}$	1.5
Sodium molybdate	$56.2^{(0)}$	$115.5^{(100)}$	5.5

The superscript figures are the temperatures (°C) at which the solubilities were determined.

of ammonium nitrate where it is desired to reduce the amount of nitrate nitrogen present. The weights of fertilisers required to make one litre of stock solution which then has to be diluted at 1 in 200, i.e. one part of stock solution made up to 200 parts with water, to give feeds ranging in strength from 50 to 300 ppm of N and K are given in Table 9.3. For practical purposes, the phosphorus level of these feeds has been fixed in relation to the nitrogen at a rate of 15 ppm P to each 100 ppm N. When potassium nitrate is used as the source of potassium, the relatively high nitrogen content of the fertiliser makes it impossible to prepare feeds having a high K and low N level. If it was intended to make a feed having 300 K and 50 N ppm, it would be necessary to use another source of K which does not contain any N, e.g. KCl or a suitable grade of K_2SO_4.

182 Liquid Feeding

Table 9.3. *Weight of fertilisers required to prepare stock solutions for dilution at 1 in 200.*

Grams per litre

Fertiliser	N 50 P 7·5	N 100 P 15	N 150 P 22·5	N 200 P 30	N 250 P 37·5	N 300 P 45	
Ammonium nitrate	16·0	42·3	68·9	95·4	121·8	148·3	
Monoammonium phosphate	6·1	12·3	18·4	24·5	30·6	36·7	K 50
Potassium nitrate	26·3	26·3	26·3	26·3	26·3	26·3	
(EC)	(0·32)	(0·57)	(0·80)	(1·04)	(1·28)	(1·51)	
Ammonium nitrate	5·4	31·9	58·4	84·8	111·3	137·8	
Monoammonium phosphate	6·1	12·3	18·4	24·5	30·6	36·7	
Potassium nitrate	52·6	52·6	52·6	52·6	52·6	52·6	K 100
(EC)	(0·41)	(0·65)	(0·89)	(1·12)	(1·35)	(1·58)	
Ammonium nitrate	*	21·4	47·8	74·3	100·8	127·3	
Monoammonium phosphate	6·1	12·3	18·4	24·5	30·6	36·7	
Potassium nitrate	78·9	78·9	78·9	78·9	78·9	78·9	K 150
(EC)	(0·58)	(0·73)	(0·96)	(1·20)	(1·43)	(1·66)	
Ammonium nitrate		10·8	37·3	63·8	90·2	116·7	
Monoammonium phosphate		12·3	18·4	24·5	30·6	36·7	
Potassium nitrate		105·3	105·3	105·3	105·3	105·3	K 200
(EC)		(0·82)	(1·04)	(1·29)	(1·52)	(1·74)	
Ammonium nitrate		Nil	26·8	53·3	79·8	106·2	
Monoammonium phosphate		12·3	18·4	24·5	30·6	36·7	
Potassium nitrate		131·5	131·5	131·5	131·5	131·5	K 250
(EC)		(0·90)	(1·11)	(1·37)	(1·60)	(1·84)	
Ammonium nitrate		†	16·3	42·8	69·2	95·7	
Monoammonium phosphate		12·3	18·4	24·5	30·6	36·7	
Potassium nitrate		157·8	157·8	157·8	157·8	157·8	K 300
(EC)			(1·22)	(1·44)	(1·69)	(1·87)	

* = 58·9 ppm N
† = 117·8 ppm N

This table was computed using the following fertiliser analysis: ammonium nitrate 35% N; monoammonium phosphate 12% N, 24·5% P; potassium nitrate 14% N, 38% K. The electrical conductivity (EC) values of the *diluted* fertiliser solutions have been expressed in millimhos at 25°C.

The weights of fertiliser required to prepare the feeds in Imperial measure, i.e. ounces per gallon, are given in Table 9.4, and in ounces per US gallon in Table 9.5.

FORMULAE FOR CALCULATING LIQUID FEEDS

Growers, who wish to use liquid feeds of different analyses or composition to those already given in Table 9.3 (or 9.4 and 9.5), can quite easily make up

Table 9.4. *Weight of fertilisers in ounces to make 1 gallon (Imperial) of stock solution for dilution at 1 in 200 to give the following liquid feeds.*

Ounces per Imperial gallon

Fertiliser	N 50 P 7·5	N 100 P 15	N 150 P 22·5	N 200 P 30	N 250 P 37·5	N 300 P 45	
Ammonium nitrate	2·56	6·78	11·05	15·30	19·53	23·78	
Monoammonium phosphate	0·98	1·97	2·95	3·93	4·91	5·89	K 50
Potassium nitrate	4·22	4·22	4·22	4·22	4·22	4·22	
Ammonium nitrate	0·87	5·12	9·37	13·60	17·85	22·09	
Monoammonium phosphate	0·98	1·97	2·95	3·93	4·91	5·89	K 100
Potassium nitrate	8·43	8·43	8·43	8·43	8·43	8·43	
Ammonium nitrate	*	3·43	7·67	11·91	16·16	20·41	
Monoammonium phosphate	0·98	1·97	2·95	3·93	4·91	5·89	K 150
Potassium nitrate	12·65	12·65	12·65	12·65	12·65	12·65	
Ammonium nitrate		1·73	5·98	10·23	14·46	18·71	
Monoammonium phosphate		1·97	2·95	3·93	4·91	5·89	K 200
Potassium nitrate		16·89	16·89	16·89	16·89	16·89	
Ammonium nitrate		Nil	4·30	8·55	12·80	17·03	
Monoammonium phosphate		1·97	2·95	3·93	4·91	5·89	K 250
Potassium nitrate		21·09	21·09	21·09	21·09	21·09	
Ammonium nitrate		†	2·61	6·86	11·10	15·35	
Monoammonium phosphate		1·97	2·95	3·93	4·91	5·89	K 300
Potassium nitrate		25·30	25·30	25·30	25·30	25·30	

* = 58·9 ppm N
† = 117·8 ppm N

184   Liquid Feeding

the feeds by using the following formulae. It has been assumed that a concentrated stock solution will first be made, which will then be diluted to the correct strength by means of either a diluter, a mixer-proportioner, an injector or similar device. All these appliances work on a volume basis, which is much more practical than the weight basis. The density of specific gravity of the stock solution will vary with the fertilisers used and the concentration; it can be 15% or more greater than plain water. All calculations have, therefore, been made on the basis of a *weight* of fertiliser being applied to a *volume* of water, i.e. w/v.

In Britain, it is proposed that the conversion of agricultural weights and measures to the metric system will be completed by 1976. This system of calculation is much simpler to use than Imperial measures, once the grower is familiar with the units, but to assist those growers who may be unable to make the change to metric with their present equipment, worked examples have also been given in Imperial measures.

(1) Calculation of weight of fertiliser required to provide a given concentration (ppm) in the diluted feed.

E.g. What weight of ammonium sulphate (21% N) is required to make a stock solution which after being diluted at 1 in 200 gives a liquid feed of 100 ppm N?

A. Metric

$$\frac{\text{grams of fertiliser per litre of stock solution}} = \frac{\text{ppm of nutrient} \times \text{dilution rate}}{1\,000} \times \frac{100}{\%\text{ nutrient in fertiliser}}$$

$$\frac{100 \times 200}{1\,000} \times \frac{100}{21} = 95{\cdot}24 \text{ g}$$

B. Imperial measure

$$\frac{\text{ounces of fertiliser per gallon of stock solution}} = \frac{\text{ppm of nutrient required} \times \text{dilution}}{62{\cdot}5 \times \text{per cent nutrient in fertiliser}}$$

$$\frac{100 \times 200}{*62{\cdot}5 \times 21} = 15{\cdot}24 \text{ oz.}$$

*(For USA substitute 75·0 for 62·5)

(2) To calculate the concentration (ppm) of nutrient in the dilute feed supplied by a given amount of fertiliser.

E.g. 25 g of urea (46% N) dissolved in 1 litre of water and diluted 1 in 200. What ppm of N does this supply?

Formulating Liquid Feeds

A. Metric

$$\frac{\text{ppm in}}{\text{dilute feed}} = \frac{\text{fertiliser weight} \times 1\,000}{\text{dilution}} \times \frac{\%\text{ nutrient}}{100}$$

$$\frac{25 \times 1\,000}{200} \times \frac{46}{100} = 57 \cdot 5 \text{ ppm N}$$

B. Imperial

16 oz. of urea dissolved in 1 gall. and diluted at 1 in 200.

$$\frac{\text{ppm in}}{\text{dilute feed}} = \frac{\text{ounces fertiliser} \times \text{per cent nutrient} \times 62 \cdot 5}{\text{dilution}}$$

$$\frac{16 \times 46 \times 62 \cdot 5^*}{200} = 230 \text{ ppm N}$$

*(For USA substitute 75·0 for 62·5)

(3) To calculate the amounts of fertilisers required to make up a complete NPK feed where fertilisers supplying more than one nutrient are used.
E.g. To make up 1 litre of stock solution to be diluted at 1 in 200 to give 200 N 30 P 150 K in the dilute feed using:

    monoammonium phosphate (12% N, 24·5% P)
    potassium nitrate   (14% N, 38% K)
    ammonium nitrate   (35% N)

Using the formulae already given in (1) and (2),

(a) calculate the amount of monoammonium phosphate required to supply 30 ppm P;
(b) calculate the amount of N this also supplies;
(c) calculate the amount of potassium nitrate required to supply 150 ppm K;
(d) calculate the amount of N this also supplies;
(e) add the amounts of N supplied by the monoammonium phosphate and the potassium nitrate and subtract this from the required value of 200 ppm N;
(f) calculate the amount of ammonium nitrate required to supply this amount of N.

Metric

(a) grams of monoammonium phosphate required $= \dfrac{30 \times 200}{1\,000} \times \dfrac{100}{24 \cdot 5} = 24 \cdot 5 \text{ g}$

(b) this also supplies nitrogen $= \dfrac{24 \cdot 5 \times 1\,000}{200} \times \dfrac{12}{100} = 14 \cdot 7 \text{ ppm}$

186 Liquid Feeding

(c) grams of potassium nitrate required $= \dfrac{150 \times 200}{1\,000} \times \dfrac{100}{38} = 78 \cdot 9\,\text{g}$

(d) this also supplies nitrogen $= \dfrac{78 \cdot 9 \times 1\,000}{200} \times \dfrac{14}{100} = 55 \cdot 2\,\text{ppm N}$

(e) total N so far supplied
$= 14 \cdot 7 + 55 \cdot 2 = 69 \cdot 9\,\text{ppm N}$
$200 - 69 \cdot 9 = 130 \cdot 1\,\text{ppm N still required}$

(f) grams of ammonium nitrate required $= \dfrac{130 \cdot 1 \times 200}{1\,000} \times \dfrac{100}{35} = 74 \cdot 3\,\text{g.}$

Therefore to make this liquid feed requires

24·5 g monoammonium phosphate
78·9 g potassium nitrate
74·3 g ammonium nitrate

Imperial

(a) ounces of monoammonium phosphate required $= \dfrac{30 \times 200}{62 \cdot 5^* \times 24 \cdot 5} = 3 \cdot 9\,\text{oz.}$

(b) this also supplies nitrogen $= \dfrac{3 \cdot 9 \times 12 \times 62 \cdot 5^*}{200} = 14 \cdot 7\,\text{ppm N}$

(c) ounces of potassium nitrate required $= \dfrac{150 \times 200}{62 \cdot 5^* \times 38} = 12 \cdot 6\,\text{oz.}$

(d) this also supplies nitrogen $= \dfrac{12 \cdot 6 \times 14 \times 62 \cdot 5^*}{200} = 55 \cdot 2\,\text{ppm N}$

(e) total nitrogen supplied so far
$= 14 \cdot 7 + 55 \cdot 2 = 69 \cdot 9\,\text{ppm N}$
$200 - 69 \cdot 9 = \dfrac{130 \cdot 1\,\text{ppm of N}}{\text{still required}}$

(f) ounces of ammonium nitrate $= \dfrac{130 \cdot 1 \times 200}{62 \cdot 5^* \times 35} = 11 \cdot 9\,\text{oz.}$

In Imperial measure, this feed requires

3·9 oz. monoammonium phosphate
12·6 oz. potassium nitrate
11·9 oz. ammonium nitrate

in 1 gall. of water, dilute at 1 in 200.

* (For USA substitute 75·0 for 62·5.)

(4) Proprietary fertilisers.

A. Calculation of the nutrients present when a proprietary fertiliser is diluted. E.g. A proprietary fertiliser in concentrated form contains 6% N and is diluted at 1 in 200; what will be the amount of N in the diluted feed?

$$\text{ppm in feed} = \frac{\% \text{ nutrient (w/v) in concentrate} \times 10\,000}{\text{dilution}}$$

$$\frac{6\% \text{ N} \times 10\,000}{200} = 300 \text{ N}$$

B. The concentrated fertiliser contains 8% (w/v) N; what dilution is required to give 200 ppm?

$$\text{Dilution required} = \frac{\% \text{nutrient (w/v)} \times 10\,000}{\text{ppm required}}$$

$$\text{Dilution} = \frac{8 \times 10\,000}{200} = 400$$

i.e. dilute at 1 in 400.

USA MEASURES AND PRACTICE

The US gallon is approximately equal to 0·83 or $\frac{5}{6}$ of an Imperial gallon and the amount of fertilisers used in making up liquid feeds must be adjusted accordingly; more precise conversions of US to metric and US to Imperial liquid measures are given in Tables 1 and 2 in the Appendices (p. 260). The necessary alterations in the weights of fertilisers required to make up the complete range of stock solutions based on ammonium nitrate, monoammonium phosphate and potassium nitrate are given in Table 9.5. The salinity of the feeds will, of course, be the same as for the corresponding feeds in Table 9.3. Some models of Solu Bridge conductivity meters measure the EC in units of 10^{-3} whilst other models are calibrated in units of 10^{-5}. To convert from one unit to another move the decimal point two places, i.e. an EC reading of 0.75×10^{-3} measured on a recent model, is equivalent to 75×10^{-5} measured on an older model.

It is also possible for growers in the USA to make up concentrated stock solutions from water soluble fertilisers having a range of N P K ratios, and examples of their use are given in Table 9.6.

Calcium nitrate is used in nursery mixed feeds in the USA to a much greater extent than in Britain. The calcium in the fertiliser prevents the pH of the compost from falling too low and the nitrogen is all in the nitrate form. A feed having all the nitrogen in the nitrate form can be made from:

188 *Liquid Feeding*

	Ounces per US gallon
Potassium nitrate	14·75
Calcium nitrate	21·00

dilute at 1 in 200 to give 200 N 200 K ppm.

In Britain the calcium present in most water supplies usually results in the pH of the compost rising rather than falling (Table 9.7).

Calcium nitrate must not be used in a stock solution containing a complete range of macro- and micro-elements, otherwise precipitation will occur.

Table 9.5. *Weight of fertilisers in ounces to make 1 US gallon of stock solution for dilution at 1 in 200 to give the following liquid feeds.*

Fertiliser	N 50 P 7·5	N 100 P 15	N 150 P 22·5	N 200 P 30	N 250 P 37·5	N 300 P 45	
Ammonium nitrate	2·14	5·65	9·20	12·74	16·26	19·80	
Monoammonium phosphate	0·81	1·64	2·46	3·27	4·09	4·90	K 50
Potassium nitrate	3·51	3·51	3·51	3·51	3·51	3·51	
Ammonium nitrate	0·72	4·26	7·80	11·32	14·86	18·40	
Monoammonium phosphate	0·81	1·64	2·46	3·27	4·09	4·90	K 100
Potassium nitrate	7·02	7·02	7·02	7·02	7·02	7·02	
Ammonium nitrate	*	2·86	6·38	9·92	13·46	17·00	
Monoammonium phosphate	0·81	1·64	2·46	3·27	4·09	4·90	K 150
Potassium nitrate	10·54	10·54	10·54	10·54	10·54	10·54	
Ammonium nitrate		1·44	4·98	8·52	12·04	15·58	
Monoammonium phosphate		1·64	2·46	3·27	4·09	4·90	K 200
Potassium nitrate		14·06	14·06	14·06	14·06	14·06	
Ammonium nitrate		Nil	3·58	7·12	10·65	14·18	
Monoammonium phosphate		1·64	2·46	3·27	4·09	4·90	K 250
Potassium nitrate		17·56	17·56	17·56	17·56	17·56	
Ammonium nitrate		†	2·18	5·72	9·24	12·78	
Monoammonium phosphate		1·64	2·46	3·27	4·09	4·90	K 300
Potassium nitrate		21·07	21·07	21·07	21·07	21·07	

* = 58·9 ppm N
† = 117·8 ppm N

Table 9.6. *Preparation of liquid feeds from complete N P K water soluble fertilisers.*

Based on US gallons

Fertiliser			Equivalent			Ounces fertiliser per US gallon diluted 1 in 200	ppm of		
N	P_2O_5	K_2O	N	P	K		N	P	K
20	20	20	20	8·8	16·6	26·6	200	88	166
15	30	15	15	13·2	12·5	35·5	200	176	167
14	14	14	14	6·2	11·6	38·0	200	88	166
21	7	7	21	3·1	5·8	25·3	200	30	55
20	5	30	20	2·2	24·9	26·6	200	22	249
25	10	10	25	4·4	8·3	21·3	200	35	66

To obtain 100 ppm of N use half the above weights of fertiliser, and for 150 ppm N use three-quarters of the weight.

Under these conditions, the calcium nitrate must either be applied by a separate diluter or by using a twin injector, e.g. the Smiths 'Measuremix' injector. In the USA some of the samples of calcium nitrate produce a scum when dissolved to make up the concentrate. This is caused by the paraffin wax and oil used in its manufacture as conditioning agents to reduce water vapour absorption. A wetting agent can be used to emulsify the paraffin and allow it to pass through the injector.

LIQUID FEED STRENGTH AND SALINITY

Liquid fertilisers are dilute salt solutions and therefore have a salinity value. For each liquid feed, irrespective of whether it is based on a single salt or a mixture of salts, there is a direct relationship between the strength of the solution and its salinity value. The salinity of a liquid feed is most easily determined from its electrical conductivity (EC) by means of a small conductivity meter. Once this value is known, it can be used to monitor the strength of the feed being applied. The EC values of the feeds given in Table 9.3 have been included in parentheses; these values are for the *diluted* feeds. It must be remembered that the conductivity value of the plain water used to dilute the concentrated feed must be *added* to the values in the table. For example, the 200 N, 30 P, 150 K ppm feed has an EC value of 1·20 millimhos. The public water supply at the Glasshouse Crops Research Institute has an EC of 0·48 millimhos, whilst water from the Institute's borehole has an EC of 0·75 millimhos. Consequently, the EC value of this particular feed will be 1·68 millimhos if the public water supply is used and 1·95 millimhos if the bore hole water is used.

For practical purposes, the EC value of a liquid feed can be calculated from the separate values of the various fertilisers, once these are known. This

190 *Liquid Feeding*

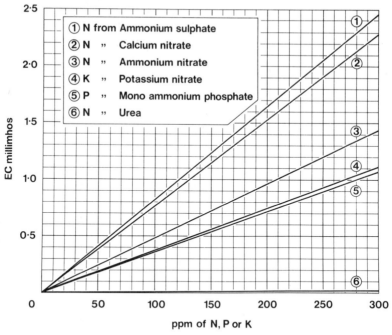

Fig. 9.2 Relationship of salinity of the liquid feed to the amount of nitrogen, phosphorus or potassium supplied by different fertilisers. Nitrogen supplied by ammonium sulphate, calcium nitrate, ammonium nitrate, urea. Phosphorus supplied by monoammonium phosphate. Potassium supplied by potassium nitrate.

relationship has been determined for a number of fertilisers commonly used to prepare liquid feeds and is given in Figure 9.2. An example of using this method to compute the EC value of a liquid feed is as follows:

What is the EC of a feed, having 200 N, 30 P, 150 K ppm, made from ammonium nitrate, monoammonium phosphate and potassium nitrate?

	Fertiliser solution		*EC value*
First,	the potassium nitrate at 150 ppm K (this also supplies 55 ppm N)	=	0·55
Second,	the monoammonium phosphate at 30 ppm P (this also supplies 15 ppm N)	=	0·11
Third,	the remaining 130 ppm of N (supplied by the ammonium nitrate)	=	0·62
	Total EC	=	1·28

To this must be added the EC of the water supply.

If desired, the ratio of ammonium to nitrate in the above feed could be changed by using 56·6 g of urea (46% N) in place of the 74·3 g of ammonium nitrate (35% N). The feed will then contain 72% of NH_4–N and 28% NO_3–N in place of the previous ratio of 40% NH_4–N and 60% NO_3–N. The EC of the diluted feed would then be 0·73 millimhos, because the urea in solution does not ionise (Fig. 9.2).

PHOSPHORUS IN LIQUID FEEDS

When pot plants were grown in composts based on mineral soils, phosphorus was frequently omitted from the liquid feed. This was because most mineral soils are able to retain a large amount of the added phosphorus and also have reserve forms of phosphorus present. In most cases, the addition of superphosphate in the base fertiliser was sufficient to meet the plant's needs. The much greater water-solubility of phosphorus in loamless composts and the increased risk of its loss by leaching has been examined in Chapter 4. It is desirable, therefore, when using loamless composts, to include some phosphorus in the liquid feed. Often this does not present any difficulties, but with some water sources a precipitate can form which will block trickle systems of irrigation.

Problems associated with the composition and use of liquid feeds have been investigated in the Horticulture Department at the Glasshouse Crops Research Institute. This work has shown that the precipitation of phosphorus in liquid feeds is dependent upon a number of factors; the quality of the phosphorus source and the quality of the water, i.e. its pH, calcium and bicarbonate contents being the most important. In Britain, the most suitable form of phosphorus for making up liquid feeds on the nursery is monoammonium phosphate. It is essential, however, that the high purity monoammonium phosphate made from 'thermal grade' phosphoric acid is used and not a lower grade fertiliser made from 'wet process' phosphoric acid. A typical analysis of the high grade material is:

N 12·1, P 26·6, Fe_2O_3 0·03, Al_2O_3 0·03, CaO 0·04, MgO 0·02, F 0·2, SO_3 0·3, SiO 0·00, As 0·0003%.

Neither superphosphoric acid nor ammonium polyphosphates are as yet generally available to the grower in Britain.

A typical stock solution containing monoammonium phosphate, which has to be diluted at 1 in 200 before application, has a pH of about 4·5 and can be kept for an indefinite period without any precipitation occurring. When the stock solution is diluted at 1 in 200 to give a liquid feed of 200 N, 30 P, 150 K ppm, the pH rises to about 7, and, depending upon the quality of water supply, a white gelatinous precipitate of iron and aluminium phosphates can form within 24 hours. This precipitate can eventually cause blockages in trickle or drip irrigation nozzles. With most water sources almost all of the phosphorus will be still in solution, however, and will be precipitated out at only a slow rate over several days, the actual rate depending principally upon the water quality and the temperature

Where the combination of the local water supply and the phosphorus source would otherwise lead to the formation of the white gelatinous precipitate, this can be prevented by the addition of the disodium salt of EDTA at 15 ppm in the dilute liquid feed. This will also prevent the formation of calcium and magnesium phosphates which can occur with some very 'hard' waters. The use of polyammonium phosphates in place of monoammonium phosphate has not, however, been successful in sequestering the iron and aluminium and preventing the precipitate from forming.

MAGNESIUM IN LIQUID FEEDS

Magnesium can be included in the liquid feed for such plants as tomato and *Solanum capsicastrum* at the rate of 15 ppm Mg in the dilute feed. This can be obtained by adding magnesium sulphate (Epsom salts, 9·9% Mg) to the stock solutions at the following rates:

$$\left. \begin{array}{l} 4\cdot9 \text{ oz./Imperial gall.} \\ 4\cdot0 \text{ oz./US gall.} \\ 30.3 \text{ g/l} \end{array} \right\} \text{ dilute at 1 in 200}$$

At this concentration there will be no problem with phosphorus precipitation.

9.3 PRACTICAL ASPECTS OF FEEDING

DISSOLVING FERTILISERS

When preparing stock solutions, it is convenient to use either hot water to dissolve the fertilisers, or to heat the cold water with a steam hose after the fertilisers have been added. Potassium nitrate causes a significant drop in the temperature of the water as it dissolves, it also has a relatively low solubility in cold water (Table 9.2).

FERTILISER DYES

It is essential, when using displacement type diluters, that the stock solution be coloured to distinguish it from the plain water which is entering the diluter. Dyeing the stock solution enables the user to see whether the stock solution has been used up and also whether any mixing of the two liquids is occurring within the diluter. At certain times, it is also useful to be able to see from the weak colour of the liquid being discharged from the end of the irrigation line that the diluter is supplying liquid feed and not plain water. *The colour of the solution cannot however be used to check the accuracy of the diluter, this can only be done by using a conductivity meter (p. 193).* Dyes suitable for colouring the stock solution include Disulphine Blue VN 150 (lissamine turquoise VN 150), fluorescein sodium (uranin) and naphthalene orange G 125 (lissamine orange G 125). They are used at the rate of 1 ounce to 20 gallons of stock solution (1 g per 3 l). It is useful to reserve a separate colour dye for each feed; this provides a ready means of identifying the feed being used, and if white plastic tanks are used to store the stock solutions the amount of feed remaining in the tanks can be seen readily without removing the lids. When using positive displacement type injectors, it is not essential to dye the stock solution.

CONTROL OF LIQUID FEED STRENGTH

The strength at which the liquid feeds are applied to the plants can be varied in two ways. One method, as shown in Table 9.3, is to make up stock solutions of different strengths and apply them at a fixed dilution rate of 1 in 200. An alternative method is to make up a stock solution of fixed strength and to vary the rate at which it is diluted or injected. Using a dilution rate of 1 in 100 would double the strength of the feed whilst a rate of 1 in 400 would halve the strength. Whilst this method appeals because of its simplicity, it is subject to difficulties. If some stock solutions are increased in strength to allow for a higher dilution rate, there may be trouble over 'salting-out' at low temperatures. Equally, if very weak solutions are made with the intention of using low rates of dilution, say 1 in 50 or 1 in 100, the relatively low specific gravity of the stock solution will mean that the risk of its mixing with the incoming displacement water is much greater, and accuracy of dilution is lost. If it is desired to use these low rates of dilution, either the two liquids must be physically separated (p. 199) or a positive type injector should be used. Also, with some diluters it is either difficult or impossible to change the dilution rate. For these reasons the constant dilution rate at 1 in 200 is preferred.

It has already been seen that each liquid feed has its own salinity or EC value. Once this is known, a small inexpensive conductivity meter can be used to check on the accuracy of the diluting equipment. The determination is easily made by placing a sample of the *diluted* liquid feed in the cell of a direct reading conductivity meter. With some meters, it is necessary to set the meter to the temperature of the liquid, other models have an automatic temperature compensation. It must be remembered that the EC of the plain water must be *added* to the EC values of the feeds given in Table 9.3 and Figure 9.2. If feeds are made up with chemicals other than those used in Table 9.3, all that is required to determine the correct EC of the diluted feed is to accurately make a dilution at the same rate as the diluter is set to operate. Flow-type electrodes permanently fitted into the pipe leading from the diluter enable quick checks to be made whilst the equipment is in operation. Suitable conductivity meters are made by Electronic Switchgear Ltd, Hitchin, Herts, England, and the Volmatic Co., Dunstable, Beds, England. The Solu Bridge is manufactured by Beckman Instruments Inc., 89 Commerce Road, Cedar Grove, New Jersey, USA.

CORROSIVE PROPERTIES OF STOCK SOLUTIONS

Whilst concentrated solutions of some fertilisers can be corrosive, those formulations made up by the grower from the materials at his disposal do not usually cause any serious trouble and the proprietary liquid fertilisers are manufactured under close control. Hatfield *et al.* (1958) examined the corrosive effect of liquid fertilisers on various metals and concluded that both stainless steel and mild steel were satisfactorily resistant to corrosion. Whilst some aluminium alloys were found to be satisfactory, others, however, were not. Solutions with a $N:P_2O_5$ ratio of 1:3 were three times as corrosive as solutions with a 1:1 ratio. Potassium was found to give a marked reduction in the amount of corrosion, even when potassium chloride was used. In many instances, the addition of 0·1% of sodium dichromate to the solutions decreased corrosion by 90%. Ammonium thiocyanate is also used as a corrosion inhibitor. Where an acid is applied through an injector to reduce the compost pH or to clear blocked irrigation lines, only phosphoric acid can be safely used because of the corrosion risk.

pH CONTROL WITH LIQUID FEEDS

It has already been shown in Figure 9.1 that liquid fertilisers can have very large effects upon the salinity of the compost, but they can also significantly affect its pH. In areas where the water supply contains large amounts of calcium and magnesium carbonates, the frequent irrigation of pots can result in an appreciable rise in the pH. For example, at the Glasshouse Crops Research Institute, where the public water supply has an alkalinity equivalent to 188 ppm calcium carbonate, the pH of the compost can increase by 1·0 to 1·5 units over the course of 10 weeks in the summer when pots are being watered twice per day. In winter, when there is less frequent irrigation, the rise in pH is much less. Quite large pH gradients can also develop in the pot; in some cases the top 3 cm layer of compost can be 1·8 pH units higher than the lower 3 cm layer (Bunt, 1956). Such increases in compost pH can create nutritional problems and there are several ways of treating the irrigating water or of preventing the compost pH from rising:

(A) By passing the water through ion exchange columns which remove all the cations and anions. The resins must be recharged after a period, and this method is much too costly except in cases where only relatively small volumes of water are required, i.e. for mist propagation units.
(B) Neutralising the water with an acid. Both sulphuric and nitric acids have been used for this purpose but neither can be recommended for reasons of safety. Phosphoric acid is a much safer material to handle but it is

important to use the correct grade. The 'wet process' acid may contain significant amounts of arsenic, and arsenical toxicity has been reported in azaleas and carnations. In Britain, the laboratory grade acid has a purity of 88–93% and does not contain any significant amounts of toxic elements, whilst in the USA the 75% food grade acid is used for water neutralisation. The quantity of acid required will, of course, depend upon the amount of bicarbonates present in the water. For example, water having an alkalinity equivalent to 2 milliequivalents of bicarbonates per 1 000 litres will require approximately: (a) 80 ml of phosphoric acid per 1 000 litres, or (b) 2·4 fl.oz. of acid per 200 imperial galls or (c) 2 fl.oz. of acid per 200 US galls. This quantity of acid will also supply approximately 30 ppm of phosphorus to the water; the precise value will depend upon the grade or purity of the acid used. When it is proposed to treat the irrigation water in this way, an advisory chemist should be consulted as to the precise amount of acid to use.
(C) The use of acidifying fertilisers in the liquid feed, of which ammonium sulphate is the best known.

Table 9.7. *Control of compost pH of pot chrysanthemums by liquid fertiliser formulation.*

Treatment	Compost pH	
	M/100 $CaCl_2$	Water
pH at start of experiment	5·58	6·15
pH at finish:		
Plain water	6·78	7·35
Ammonium nitrate		
Monoammonium phosphate		
Potassium nitrate with ammonium sulphate		
added at:		
Nil ppm of N	6·63	6·83
32·5 ppm of N	6·25	6·42
65·0 ppm of N	5·59	5·89
97·5 ppm of N	4·79	5·20
130·0 ppm of N	4·08	4·38
Ammonium nitrate ⎫ Monoammonium phosphate ⎬ Potassium sulphate ⎭	5·81	6·05
Calcium nitrate ⎫ Monoammonium phosphate ⎬ Potassium nitrate ⎭	6·98	7·31

pH determinations made on a summer crop of chrysanthemums 10 weeks after potting.
All feeds contained 200 N 30 P 150 K ppm and were made with bore hole water containing 250 ppm equivalent calcium carbonate.
In winter, pH changes due to water quality and fertiliser composition are less marked.

Liquid Feeding

The actual amount of ammonium sulphate required to control the pH will depend upon several factors, two of the most important being the total alkalinity of the water and the extent to which the compost is buffered against pH change. Results from an experiment, where a number of liquid feeds having the same NPK concentration but different potential acidities, are given in Table 9.7. The water used to make the liquid feed had an alkalinity equivalent to 250 ppm $CaCO_3$, and the buffer capacity of the compost was similar to the Irish peat–sand mixture shown in Figure 4.7 (p. 73). Under these conditions, ammonium sulphate added to the liquid feed at a rate equivalent to 25–50 ppm of nitrogen will prevent the natural increase in the pH occurring and will also improve the colour of chrysanthemum leaves. If ammonium sulphate is used at rates above 50 ppm of N in the liquid feed, there is the risk of a substantial increase in the compost salinity.

Domestic water softeners, which replace the calcium in the water with sodium, are not suitable for treating water that is to be used for irrigating plants.

RATES OF FEEDING

Whilst liquid feeding is generally regarded as an essential part of pot plant culture, experimental data upon which to base the composition and frequency of feeding are often lacking. Growers must then apply liquid fertilisers on an empirical basis.

General principles used to determine the strength of the feeds are: (1) the vigour of the plant and its size in relation to the volume of compost in which it is growing, (2) whether slow-release fertilisers have been used, (3) the season, and (4) the extent to which nutrient losses occur by leaching. The composition of the feed is often determined by the known or believed changes which occur in the nutrient balance within the plant with age and season.

When continuous feeding is practised, the nitrogen level of the feed for most pot plants is maintained within the 100–200 ppm range, with the potassium being adjusted in relation to the plant and the season. Often the nitrogen level is reduced and the potassium level increased as the plant approaches maturity.

The possible increase in the salinity of the compost toward the end of the crop, because of the increased frequency and amount of liquid fertiliser being applied, has already been discussed on page 177. To meet this situation, the feed should be either reduced to half strength or given only at alternate irrigation as the plants reach maturity. Where the liquid feeding and irrigation system is designed in such a way that one diluter is used to feed plants at all stages of development, such selective feeding is impossible. Obviously the irrigation system must be sufficiently flexible in design to allow different strengths of feed and even plain water to be given when necessary.

Some examples of liquid feed strengths, and their variation with the

Table 9.8. *Liquid feed strengths for pot plants (ppm in diluted feed).*

Species	N	P	K	Remarks
Chrysanthemum				
Summer	200	30	150 ⎫	Surface watered by hand
Winter	150	22	200 ⎭	or system drip
Summer	150	22	150 ⎫	Capillary bench or
Winter	100	15	200 ⎭	sub-irrigation system
Hydrangea	100–200	15–30	100–200	For growing
	50–100	5	200	Forcing blue varieties
Poinsettia	250	30	150	Plus 0·1 ppm of molybdenum
Azalea	50–150	10–20	20–50	Ammonium preferred to nitrate nitrogen. Maintain pH at 5·0
Foliage plants	100–200	15–30	100–200	
Saintpaulia	50–100	7–15	50–100	Nitrate nitrogen preferred. Use low N for variegated forms
Tomato	200	30	350	High potassium is necessary for fruit quality. Often 15 ppm Mg is included.

species, the season and the type of irrigation system, are given in Table 9.8. Until more experimental data is available on the requirements of different crops, the grower must be guided largely by experience and the appearance of the plants.

9.4 DILUTING EQUIPMENT

Dilute fertiliser solutions suitable for applying to pot plants can be obtained by two means. Either the required weights of fertilisers are dissolved in a measured volume of water and the solution is then ready for applying directly to the plants, or a highly concentrated stock solution is first prepared and this is diluted before use. The former method has the advantage of a high degree of accuracy and the avoidance of problems with the precipitation of certain elements which can occur in concentrated solutions. The principal disadvantage of this system is the storage of large volumes of dilute fertiliser solution. This method is, therefore, used only in certain special cases. With the range of equipment now available, the dilution of small volumes of highly concentrated stock solutions into a low-strength liquid feed is by far the most popular method of applying liquid solutions. Diluting equipment can be classified into two groups, those which operate on the displacement principle and those which operate by positive injection.

198   *Liquid Feeding*

DISPLACEMENT DILUTERS

These are relatively simple and inexpensive pieces of equipment with no moving parts; the operating principle will be clear from Figure 9.3. The container is first filled with the stock solution, to which some dye has been added, and is then connected into the water supply line. The rate at which water from the high-pressure inlet side is allowed to enter the diluter, and thereby displace an equal volume of stock solution into the water flow line on the low-pressure or down-stream side of the diluter, is controlled by two orifices. The one on the inlet side is in the head of the diluter, the other is in the pipe connecting the stock fertiliser solution to the diluter outlet. Diluters of this type are sometimes incorrectly described as operating on the venturi principle. Diluters of the displacement type are manufactured by

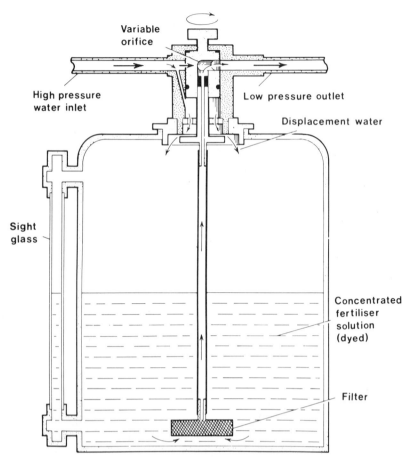

Fig. 9.3   The working principle of a displacement-type diluter.

Cameron Irrigation Co., Littlehampton, Sussex, England, and Volmatic Co., Dunstable, Beds, England.

To function satisfactorily, it is of course necessary that the incoming displacement water does not mix with the concentrated fertiliser solution and thereby dilute it. If this occurs, the effective dilution rate is reduced, resulting in a loss of accuracy. The plain water which enters at the top of the diluter has a density of 1·0, whilst the stock solution has a density of 1·15 or more. Because of these large differences in density, the plain water forms a layer on top of the high density fertiliser solution in the bottom of the diluter, and diffusion and mixing of the two solutions does not occur to any great extent. This is demonstrated by the normally sharp line of separation between the plain water and the dyed fertiliser solution as seen in the sight glass. This type of diluter should not be moved or shaken up, otherwise mixing of the two solutions will occur.

One method which prevents the mixing of the two liquids, and which also allows the diluter to be made portable, is to place the concentrated fertiliser solution into a rubber bag inside the diluter. The incoming displacement water is kept outside the bag, which slowly collapses as the fertiliser solution is displaced. Diluters employing this principle are the 'Gewa', manufactured by G. Wagner, 89 Augsberg, Hessenbachstrasse 71, Western Germany; the 'Liquid Proportioner', distributed in the UK by R.B. Controls, Goostrey, Cheshire, England; and the 'Mixer–Proportioner', distributed by Young Industries, Box 943, Los Altos, California 94022, USA.

Flow-rate control. One source of error inherent with the displacement type diluter is the effect which fluctuations in the rate of water flow have on the rate of dilution. Variations in the pressure in the water mains, or the extent to which the valve or faucet connecting the water main to the diluter is opened, can materially affect the amount of water passing through the diluter. Often, when hose watering, the operator will vary the flow rate in relation to the size of pot and this causes inaccuracies in the dilution rate. One method of obtaining a reasonable degree of control over the rate of water flow is by means of a flow control valve. This operates on the principle of a variable orifice which is controlled by the water pressure. At low pressure, a flexible rubber ring is in the open position, giving the maximum orifice size. As the hydraulic pressure in the water main increases, the ring is forced down on to a tapered seating, thereby reducing the size of the orifice and restricting the rate of water flow. Examples of constant flow-rate valves of this type are Maric Flow Valves, made by Barflo, Eastbourne, Sussex, England; and Constaflo Control, made by Laycock Engineering, Millhouses, Sheffield, England. Both of these controllers are available in a range of sizes and flow-rates and give an accuracy of $\pm 10\%$. Where different flow rates are

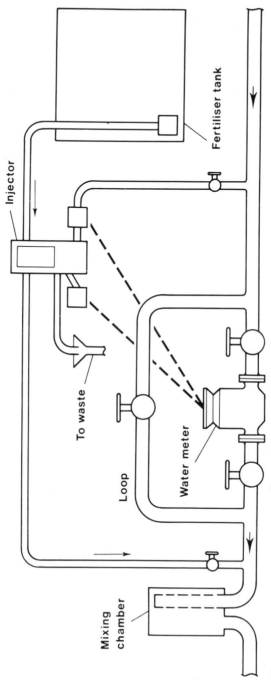

Fig. 9.4 Arrangement of injection-type diluter in the water mains, showing the water by-pass loop, which allows the diluter to be removed from the system or plain water to be given to the plants.

required when irrigating pots of different sizes, it would be necessary to use more than one size of flow control valve.

When incorrectly maintained and operated, displacement diluters can give very large fluctuations in the strengths of the liquid feed. If they are used correctly and are frequently checked with a conductivity meter, a reasonable accuracy of dilution can be achieved. These diluters have the advantage of a relatively low capital cost.

POSITIVE INJECTION DILUTERS

Positive injection diluters operate by injecting a set amount of the stock fertiliser solution into the water line. Diluters of this type can be classified into two groups, (a) those in which the injection rate is automatically controlled by the rate of water flow, and (b) those having a constant rate of injection irrespective of the rate of water flow.

Group (a) gives the greatest accuracy of dilution under nursery conditions; often it is better than 3%. They also allow the water flow-rate *to be deliberately varied* when feeding pots of different sizes and still maintain the same ratio and accuracy of injection. Diluters in this group vary somewhat in detail, but the working principle is shown in Figure 9.4. The flow of water is metered by a rotary piston type of water meter having electrical contacts built into the head. The contacts are wired to two solenoid valves fitted on the inlet and outlet sides of the injector which is operated by hydraulic pressure from the water mains. At each stroke, a fixed amount of concentrated fertiliser solution is injected into the main water flow line; the small amount of water used to operate the injector is run to waste. The speed at which the injector operates is controlled by the rate of water flow. Different rates of dilution can, however, be obtained either by altering the frequency at which the electrical impulses are transmitted from the head of the water meter to the injector, or by varying the length of the injector stroke and thereby the volume of fertiliser injected. Injectors of this type give a high degree of accuracy, providing that there is no wear on the piston washers or corrosion of the cylinder. Also, the rate of flow through the meter must not exceed its rated capacity and the pressure of water used to operate the injector must not drop below 35 psi. With this type of injector, the fertiliser solution is not injected continuously but as regular pulses. If there is a sufficient length of delivery pipe between the injector and the point of discharge, with the usual number of fittings such as elbows and T-pieces which create some turbulence in the water flow, the fertiliser will have mixed uniformly with the water. If the delivery pipe is too short to allow adequate mixing to occur, then a mixing chamber is necessary.

Several other types of positive injection dilutors are available; some operate entirely by water pressure and do not require electricity, others use an electrically-driven pump. When the irrigation water contains fine particles of sand which could score the cylinders, it is advisable to use a diaphragm type of injector.

Examples of positive injection diluters controlled by the rate of water flow are:

(1) HPA Metering Systems: T. H. Baggaley Ltd, 34 Rothschild Street, London S.E. 27, England.
(2) Commander Proportioner: Merit Industries Inc., P.O. Box 8075, Cranston, Rhode Island, 02920, USA.
(3) Fert-O-Jet: Fert-O-Jet Co., 4701 Old San José Road, Santa Cruz, California 95060, USA.
(4) Hydrocare Injector: Hydrocare Products Inc., Northfield, Illinois, 60093, USA.
(5) Smith's Measuremix: Smith Precision, 1299 Laurence Drive, Newbury Park, California, 91320, USA.

In Group (b), where the rate of injection is not controlled by the rate of water flow, micropumps with mechanically actuated diaphragm heads are used in industry to inject small quantities of fluids into flow lines. These pumps have the advantage of being able to handle a wide range of corrosive liquids. By fitting a number of heads to one pump, several different chemical solutions, which could not otherwise be mixed in the same container because of the risk of precipitation, can be injected separately into the water flow line. The principal disadvantage of using these pumps as liquid fertiliser injectors is that the rate of injection is not directly controlled by the rate of water flow. They can be accurately used only where there is a fixed rate of water flow and also a common electrical control of both the micro-pump and the pump controlling the water supply.

USE OF DILUTERS

Most local authorities stipulate that diluters must not be coupled directly into the public water supply. A break must be made in the supply line to guard against the possibility of chemicals siphoning back into the public water supply should there be a reduction in the mains water pressure. Often the installation of check valves in the water main is not regarded by the authority as giving an adequate safeguard against siphoning, and a water storage tank fitted with a ball float control will be required.

In the interest of economy, growers sometimes use a single large central diluter to supply liquid feed to the whole nursery. This has the disadvantage of not being sufficiently flexible to permit different feeds to be used for different plants. In some cases, it is impossible even to obtain a plain water supply in the glasshouse, only liquid feed is available. As already shown, excessive feeding can cause high salinities and thereby restrict plant growth; the ability to apply plain water to plants if required is highly desirable. A water loop to bypass the diluter should always be fitted into the system (Fig. 9.4). Not only does this enable plain water to be used if required, in the event of failure the diluter can easily be removed from the system for repair.

9.5 QUALITY OF IRRIGATION WATER

DISSOLVED SALTS

In addition to carbonates of calcium and magnesium, irrigation water may contain other salts, the most important of which, from the pot plant grower's point of view, are sodium bicarbonate and boron. Public water supplies in England do not usually contain significant amounts of either sodium or boron but, in some areas, high concentrations of these elements can occur, e.g. in western parts of the USA. In England, water from boreholes situated within three miles from the coast will often contain more sodium and chlorides than are present in the public water supply, especially in times of low rainfall when the underground seepage is greatest. Boreholes sunk deep into the chalk strata will often have large amounts of calcium and magnesium carbonates. Wall and Cross (1943) found that plants responded differently to salts in the irrigation water. Carnation and primula grew well with 1 000 ppm of sodium bicarbonate in the water, whilst tomatoes and chrysanthemums were affected by 500 ppm. The contrast in the response of carnations and poinsettias to various salts in the irrigation water is seen from the following results, the salts being arranged in order of decreasing toxicity:

Carnation, $MgCl_2 > CaCl_2 > NaCl > MgSO_4 > Na_2SO_4 > NaHCO_3 > CaSO_4$

Poinsettia, $NaHCO_3 > MgCl_2 > NaCl > CaCl_2 > MgSO_4 > CaSO_4$

When toxicity was due to sodium bicarbonate, the effect of the high pH exceeded the toxic effects of a high ionic concentration. With other salts, the toxicities were due to osmotic effects and nutrient unbalance. Sodium bicarbonate was recognised as being more toxic than sodium chloride. These workers concluded that water containing 200 ppm of bicarbonates will seldom cause toxicity, and 500 ppm is the maximum level of bicarbonate that can be used without treatment of the water. In cases where there is no practical alternative to using water containing large amounts of sodium bicarbonate, neutralisation by treatment with an acid is recommended; sulphuric and nitric acids have given better results than phosphoric acid.

Waters *et al.* (1972) have made recommendations regarding the quality of irrigation water suitable for ornamentals (Table 9.9).

Whilst plants vary in their tolerance to individual elements, for general purposes the irrigation water should not contain more than:

Ca, 100 ppm; Na, 50 ppm; Mg, 25 ppm; Chloride, 70 ppm; Sulphate, 250 ppm; Carbonates, 60 ppm; and Boron, 0·75 ppm.

Table 9.9. *Suitability of water for irrigating pot plants.*

Water classification	Electrical conductance, Millimhos at 25°C	Total dissolved solids (salts), ppm	Sodium, % of total solids	Boron, ppm
Excellent	< 0·25	< 175	< 20	< 0·33
Good	0·25–0·75	175–525	20–40	0·33–0·67
Permissible	0·75–2·0	525–1 400	40–60	0·67–1·00
Doubtful	2·0–3·0	1 400–2 100	60–80	1·00–1·25
Unsuitable	> 3·0	> 2 100	> 80	> 1·25

WATER PURIFICATION

Water drawn from ponds and rivers for plant irrigation may contain bacterial and fungal pollution. This is usually more of a problem when growing food crops than ornamental crops. There are two ways in which polluted water can be treated to enable it to be safely used for irrigation purposes; one method employs chlorination of the water, whilst in the other method it is treated with ultra-violet radiation. Chlorination of water can be achieved either by adding a sodium hypochlorite solution or by producing chlorine electrolytically from sodium chloride. In the hypochlorite method, sufficient sodium hypochlorite is added to a known volume of water to give a chlorine content of between 5 and 20 ppm. If the water is stored for a period after treatment, then a chlorine concentration as low as 0·5 ppm can be effective (Adams, 1973). In the electrolytical method, chlorine is generated by the electrolysis of sodium or potassium chloride; chlorine produced by this method is said to be in a highly reactive state and a lower concentration is required. The alternative method of treating water is by passing it through an ultra-violet light source having an emission peak of 2 537 Ångström units. The bacteria are destroyed at this wavelength, but because ultra-violet radiation has a short penetration in water, the water must be treated in shallow layers. The equipment is designed to work automatically.

Chapter 10

Irrigation systems

Plants grown in pots have only a relatively small reserve of water in the compost from which to meet their requirements, and this reserve soon becomes exhausted unless replenished by frequent irrigations. Some plants may require as many as three irrigations per day in mid-summer to prevent wilting. The high cost of labour has encouraged the use of various automatic or semi-automatic systems of irrigation and the majority of commercially grown pot plants are now watered by one or other of these systems. In the following discussion of irrigation systems, it is not the intention to make a critical appraisal of proprietary equipment, but rather to discuss the general principles and characteristics of each system in so far as they relate to plant nutrition and management.

Irrigation systems can be classified into three groups:

(1) Drip systems
(2) Capillary watering
(3) Flooded benches

10.1 DRIP SYSTEMS

With this system, illustrated in Figure 10.1, a plastic hose of $\frac{1}{2}$ in. to $\frac{3}{4}$ in. ($1\frac{1}{4}$ cm–2 cm) diameter is laid along the centre of the bench. From this hose small bore capillary tubes, sometimes known as 'spaghetti' lines, are taken to each pot and held in position over the surface of the compost with either a peg or a weight. The rate at which water flows from the tubes is determined by the water pressure in the distribution mains and by the diameter and length of the small bore tubes. The tubes are usually of 0·75 mm ($\frac{1}{32}$ in.) internal diameter. If tubes with a larger internal diameter are used, they have a restriction or nozzle fitted at the discharge end to control the rate of flow; this can be removed for cleaning if a blockage occurs. Usually the rate of water flow is between 1·5 and 4 pints/hr (1–$2\frac{1}{4}$ l/hr). This is slow enough for the water to drip from the end of the tubes, hence the term 'drip system'. The frequency and duration of the irrigation can be controlled either manually

206 *Irrigation Systems*

Fig. 10.1 A drip-type irrigation system. The diameter and length of the capillary tube leading from the black plastic water main to the pot controls the rate of water discharge.

or automatically by the use of time switches or weighing devices wired to solenoid valves. These allow a set amount of water or liquid feed to be applied either at given intervals or after the predetermined weight of water has been lost from the pot by evapotranspiration.

From the cultural viewpoint the advantages of the drip system of irrigation are:

(a) The ability to control the soil moisture tension and thereby exercise some control over the rate of growth. Plants can be grown relatively wet, to obtain the maximum growth rates, and then the water content can be reduced, to produce a check to vegetative growth and so encourage a more reproductive condition, e.g. the even production of winter flowering in the pot chrysanthemum.

(b) Pots can be leached as often as required to prevent a build-up of salts in the compost.

The disadvantages of this system are:

(a) All plants receive the same amount of water and nutrients irrespective of any difference in their growth rates and water requirements. Variations that naturally occur between plants in their water requirement mean that some plants could be over-watered whilst others are under-watered.

(b) Because of the low rate of water application, i.e. between 1·5 and 4 pints/hr (1 to 2·3 l/hr), the lateral spread of water in some composts can be inadequate; one part of the pot can be dry whilst another part is wet. This situation is worse when the 'plant' consists of several rooted cuttings grown in the same pot, e.g. the pot chrysanthemum. It is most likely to occur when the compost has been allowed to dry out too much between irrigations and can be avoided by using either a nozzle which gives a fan-shaped spray over the compost surface or a water-loop hose in each pot in place of the drip discharge. The water loop hose consists of a length of black polythene tube stitched along one side and shaped in the form of a circle or loop which fits inside the rim of the pot. Water is fed into the tube and drips or seeps out through the stitching. Both the fan-jet and water-loop hose apply water over a large surface area and the risk of wet and dry patches occurring in the same pot is avoided.

With the drip system of irrigation, excessive amounts of water can easily be given either by leaving the system turned on too long or by using it too frequently. It is desirable, therefore, to choose a compost that drains freely and always has an acceptable level of aeration; this offers some safeguard against over-watering. The detrimental effects of excessive water application and the consequent lack of aeration on plant growth has already been discussed (p. 57). However, using an open structured, well-drained compost will mean that there will probably be a greater loss of plant nutrients by leaching, and closer attention to their replacement by liquid feeding will be required.

The small bore irrigation tubes can become blocked either by dirt entering the system, by the precipitation of certain salts from the liquid fertiliser, or by algal and bacterial growth within the irrigation system. Blockages can be prevented or cleared by the following measures:

(1) Fitting 200 mesh line filters on the down-stream side of the fertiliser injector.
(2) Fitting valves at the end of the supply lines to allow them to be flushed out occasionally.
(3) The addition of the disodium salt of EDTA to the liquid feed stock solution, to give a concentration of 15 ppm in the feed after dilution, will prevent the formation of the white gelatinous precipitates of iron and aluminium phosphates and also of the calcium and magnesium phosphate precipitates, that can cause blockages with some sources of irrigation water.
(4) The occasional addition of a general algicide/bactericide to the water will prevent the growth of algae and bacteria within the irrigation system. For example, 'Panacide' used at the rate of 15 ml/1 000 l (0·5 fl.oz./200 galls) of irrigation water at fourteen day intervals. Recent experience suggests that, where the lines have become blocked by bacterial growth,

hydrogen peroxide diluted at 1 in 200 can be used without causing any detrimental effect to plant growth. Regular flushing of the lines with hydrogen peroxide at a strength of 1 in 1 000 will also prevent the build up of bacterial slime. When a crop has been cleared, the lines can be soaked for 30 minutes in a general disinfectant, e.g. 'LF-10' (Amphyl) at 2%, chlorox at 10%, or a domestic chlorine bleach diluted to give 0·5% by weight of sodium hypochlorite. Flush out the pipes with clear water before re-using.

(5) If the nozzles become partially blocked either from the formation of precipitates in the liquid feed or from the evaporation of water containing calcium carbonate, a dilute acid can be used to clear the lines. Because of the corrosion risk, phosphoric acid and citric acid powder are the only acids that can be used safely with the injector systems. Phosphoric acid at 4 fl.oz./gall (2·5 ml/l) of water diluted at 1 in 200 gives approximately 100 ppm of phosphoric acid; the actual strength depends upon the purity of the acid used. Citric acid powder can be used at the rate of 2 oz./gall of undiluted fertiliser solution, and this rate can be increased to 4 ounces per gallon if necessary. Nitric acid has been successfully used with *displacement type diluters* to clear the blocked irrigation lines of tomato crops. The acid is first diluted by adding one volume to 14 volumes of water before placing in the diluter, which is set to operate at 1 in 50 to give a final dilution of 1 in 800. Nitric acid is not suitable, however, for use with injector systems because of the corrosion risk. This method should also be tested on a small batch of plants to determine their reaction, before proceeding on a large scale.

Drip systems of irrigation are offered by:

Access Ltd, Crick, Rugby, England.
Cameron Irrigation Ltd, Littlehampton, Sussex, England.
Volmatic Ltd, 18 Matthew Street, Dunstable, Beds, England.

and in the USA by:

Chapin Watermatics Inc., P.O. Box 298, 368 Colorado Ave., Watertown, NY 13601, USA.

10.2 CAPILLARY WATERING

With the capillary system of watering, pots are stood upon a layer of wet sand or other substrate, the water then flows by capillary attraction from the sand through the base of the pot and into the compost. Essentially there are two systems of capillary watering in use:

Capillary Watering 209

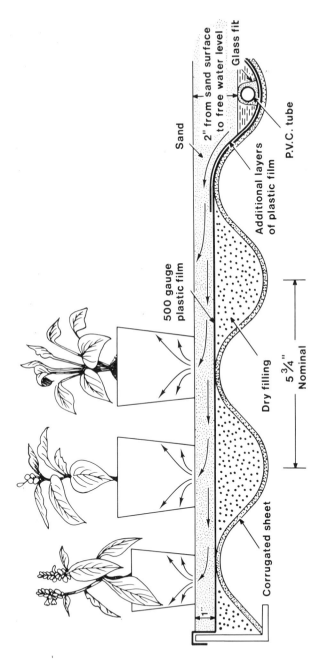

Fig. 10.2 The National Institute of Agricultural Engineering type of capillary bench.

CAPILLARY BENCHES

In the system developed by the National Institute of Agricultural Engineering (Welles and Soffe, 1962, with constructional details given in Horticultural Machinery Leaflet No. 10, Ministry of Agriculture, Fisheries and Food, Anon, 1964), a constant water-table is maintained 2 in. below the surface of a layer of sand, whose function is to distribute the water across the bench (Fig. 10.2). For this system to operate efficiently, the bench must be made level to within $\pm \frac{1}{8}$ in.; this is important as otherwise the sand in one part of the bench would be flooded, whilst in other parts it could be dry. The bench must be made waterproof by covering it with a sheet of polythene. It is also important that the sand is of the correct grade or particle size; it must be sufficiently fine to maintain capillarity and at the same time be porous enough to transmit a high rate of water flow. The suitability of the sand can be determined from two simple practical tests. First, a 3 in. diameter can with a perforated base is filled with the dry sand to within $\frac{1}{4}$ in. of the top and then placed in a shallow tray. Water is added to the outer container until it is 3 in. below the surface of the sand, which should become moist within 24 hours. For the second test, it should be possible to pour half a pint of water into the wet sand within 4 minutes without any spillage or overflow occurring.

The water level in the bench is maintained automatically by a supply tank fitted with a ball-valve. The rate at which water is lost by evaporation and transpiration is not, however, sufficiently great to work a displacement type liquid feed diluter. Liquid feeding must be done either by making up the feed on a batch system into a separate tank, from where it is fed into the supply tank as required, or it can be applied occasionally direct to the surface of the compost from a hose pipe connected to a diluter in the normal way. Any surplus water will then drain into the sand and be lost through the overflow.

IRRIGATED SAND BENCHES

The second system of capillary watering is referred to as an 'irrigated sand bench'. With this system, a layer of sand 1 to 2 in. deep is laid over a sheet of polythene and kept wet by a drip or trickle system of irrigation. No attempt is made to make the bench either level or watertight, the polythene is not turned up at the ends or edges of the bench and surplus water is allowed to drain from the bench. As there is no water reservoir from which the sand can be kept wet by capillary flow, and as the sand holds relatively little water at low tensions, it must be irrigated frequently to ensure that there is sufficient water to maintain a capillary flow to the pots. The frequency with which the sand is irrigated is usually controlled by a time-switch. Under experimental conditions a pressure-switch control, which is actuated by the amount of water in the sand, has been used successfully and offers a better means of automatically maintaining the sand in a wet condition. Control systems based on the electrical conductivity of the wet sand are not generally successful because of the variable amounts of salts present from liquid feeding.

With both systems of capillary watering, the compost is kept at a much more uniform moisture content than when pots are watered by either the drip system or by hand watering. When surface-watered, the compost can be allowed to partially dry out before re-wetting, whereas, with capillary watering, the average water tension in the pot is nominally about 5 in. or 12 cm. This means that the plants do not experience any significant water stress. Pots stood on irrigated sand benches can, however, dry out rather more than this during the warmest part of a mid-summer day if the sand is not kept sufficiently wet to maintain a water supply adequate to replace evapotranspiration losses. When the evapotranspiration rate falls during the evening, the water content of the compost rises again.

Plastic pots are the most suitable type for use on capillary benches as the sand readily makes contact with the compost through the drainage holes, thereby establishing a capillary flow. Clay pots, however, have a thick base, which prevents good contact between the compost and the wet sand, and, as the clay does not have a sufficiently high water-conductivity to supply the compost with water, a glass-fibre wick must be inserted through the drainage hole to connect the compost with the sand. The maximum height of pot which can be satisfactorily used with this system of irrigation is 6 in. (15 cm).

Recently, various materials have been tried as alternatives to the sand, these include acrylic fibre mats and a cheaper polythene-backed cotton mat. These materials have the advantage of being much lighter in weight than sand and can therefore be used on light-weight benches. As yet there is insufficient experience of their effective capillary capacity and durability to know whether they will ultimately replace the sand.

The high moisture content of the sand is also conducive to the growth of algae, especially when nitrogen and phosphorus are being supplied in the liquid feed. This algal growth can be prevented by treating the sand with algicides that are non-phytotoxic to higher plants. Two materials that have given satisfactory control of algal growth are 'Algamine' made by Gerhardt-Penick Ltd, Thornton Laboratories, Glebe Road, Huntingdon, England, and 'Panicide' made by British Drug Houses Ltd, Poole, Dorset, England.

The advantages of the capillary systems of watering from the compost and cultural viewpoints are:

(a) Water is supplied to meet only the plant's requirements and excessively high water content with low aeration is avoided. Whilst the water tension in the compost of capillary-watered pots is always low, usually being about 10 to 12 cm, the compost is nevertheless not as wet as that in pots that have just been watered overhead either by a hose pipe or a drip irrigation system. Comparable air-water contents of three composts watered by the two systems are given in Table 10.1. With the capillary system of irrigation, the air content of the composts was maintained at between two and three times the value found in surface-watered pots which had been watered and allowed to drain for 8 hours. The reason for this significantly greater air content will be apparent from Figure 3.2 (p. 48). This shows a typical compost water desorption curve and illustrates the large reduction in the water content and the corresponding increase in the air content that occurs when the compost is subjected to only a few centimetres of water tension. Thus, with the capillary system of irrigation, the air-water content remains almost constant, whereas with a surface irrigation system the air content continually fluctuates from a very low value immediately after watering to an increasingly high value as the compost dries out.

Table 10.1. *Air–water relations of three composts irrigated by different systems.*

Watering system	Compost		
	1	2	3
Hand watered (drained for 8 hours)			
Air capacity	5·4	8·0	19·1
Water	75·6	63·1	52·9
Total pore space	81·0	71·1	72·0
Capillary watered (pots 5 cm above the water table)			
Air capacity	17·1	18·0	36·6
Water	63·6	53·5	34·5
Total pore space	80·7	71·5	71·1

(b) At the constant high moisture content, the salinity stress in the compost remains relatively low and does not periodically rise to the high values associated with composts that are allowed to dry out.
(c) There is no loss of nutrients by leaching as occurs with surface watering systems.
(d) The irrigation of plants with a rosette habit of growth is much easier with this system; in particular, plants such as *Saintpaulia*, which must not have their foliage splashed with low temperature water.

The disadvantages of the capillary system of watering are:

(a) Unlike plants grown by surface watering systems, plants on capillary benches cannot be subjected to high water stress so as to produce a less vegetative and more reproductive form of growth. This is a useful characteristic of surface watering systems.
(b) The continuous upward flow of water from the sand into the compost means that fertilisers accumulate in the compost. With surface watering systems, this build-up of salts can be controlled by leaching the compost either by applying an excess of nutrient solution or, more effectively, by leaching with plain water.
(c) If the sand is not maintained sufficiently wet to give an adequate rate of water flow into the compost, the roots will grow down into the sand so that, when the pot is removed, there is extensive loss of root and possible damage to the plant.

The increase in the salinity of the compost when liquid fertilisers are added to the water is probably the most serious difficulty encountered in growing plants by the capillary systems of irrigation. Guttormsen (1969) found that, when pots of compost were stood on capillary benches for a 10 week period, significant salinity gradients developed within the pots. For example, when the irrigation water contained 200 ppm each of nitrogen and potassium, the salinity of the compost in the lower half of the pot was about 2 millimhos, whereas in the upper half of the pot it was about 9 millimhos. When the nitrogen and potassium levels were increased to 400 ppm, the salinity levels in the lower and upper parts of the pot were 3 millimhos and 20 millimhos respectively. Bunt (1973, *b*) compared the response of pot chrysanthemums grown on an irrigated sand bench with those grown by surface watering. On the irrigated sand bench, a liquid feed having 100 ppm of nitrogen gave approximately similar results, in terms of plant growth and nitrogen levels in the compost, to a liquid feed of 300 ppm of nitrogen applied to the compost surface of plants having free drainage from the base of the pot. It is often more satisfactory to apply liquid fertilisers by hand to pots stood on capillary benches than to include the fertilisers in the water supply.

Sand or capillary watering systems are at present more widely used in Britain and Holland, especially for plants grown in small pots, than in the USA where the drip system of irrigation predominates.

10.3 FLOODED BENCHES

With this system of irrigation, pots are stood on a watertight bench. Either water or a weak nutrient solution from a large reservoir under the bench is pumped up onto the bench until the pots are between one third and one half submerged. The water is left on the bench at this level until the surface of the compost in the pots becomes moist by capillary flow. The water is then allowed to drain from the bench back into the reservoir ready for the next irrigation.

This system of pot irrigation was developed by Post (1946) and Seeley (1947) at Cornell University, New York. Because water and nutrients enter the pot through the base, and as there is effectively little leaching with this system, a build-up of salts can occur, similar to that experienced with the capillary bench system.

With a recently introduced variation of the flooded-bench system, irrigation is achieved by a shallow layer of flowing water rather than by a greater depth of static water. The pots are stood on shallow, watertight benches having a fall or slope of 0·5% (about 1 in. in 16 ft). Water or a nutrient solution is pumped on to the high end of the bench from a reservoir and is distributed across the bench by a header or feeder pipe. Normally, it is released for a three minute period, at the rate of 5 galls per minute for each foot width of bench. The water flows down the bench at a rate of 20 ft per minute, flooding the pots to a depth of $\frac{3}{8}$ in. At the low end of the bench, the water is collected and drained back into the reservoir for re-use. This system of irrigation has not been in use for a sufficiently long period to give information on the possible interactions that could occur with plant nutrition. As only a very shallow layer of compost at the bottom of the pot is irrigated by free water followed by normal drainage, with the remainder of the compost being irrigated by capillary flow, it therefore seems likely that the same general principles of the capillary system of watering will also apply to this system.

Chapter 11

John Innes composts

The John Innes composts were developed as a result of investigations by Lawrence and Newell (1939), at the John Innes Horticultural Institution, into the high mortality rate of *Primula sinensis* seedlings being raised for genetical studies. This work was the first attempt by researchers to produce a standardised compost suitable for a wide range of plants. Prior to this, there had been no standardisation of materials or compost formulae; growers prepared individual composts for each species of plant from carefully preserved and guarded recipes. In addition to the well-rotted turf loam, chopped with a spade into walnut size pieces, which formed the basis of the composts, a wide range of other materials was used, including mortar rubble, crushed brick, burnt clay, charcoal, well-rotted animal manure, spent hops, leaf mould, etc., and fertilisers such as bone meal, steamed bone flour, flue dust, bonfire ashes, etc.

The John Innes composts differed from the traditional composts in a number of important ways:

(a) A single physical mixture was found to be suitable for a wide range of plants.
(b) They were based upon steam sterilisation of the soil which, if done correctly, eliminated the heavy losses of seedlings resulting from soil-borne fungal and insect attacks.
(c) A 'sterile' and relatively standard organic physical conditioning material, i.e. peat, was used, in place of rotted animal manure and other materials of variable physical and nutritional properties.
(d) Fertilisers were added with precision on a weight basis, in place of the traditional advice of '... to each barrow-full of this mixture add a six inch pot-full of bone meal, sprinkle a handful of lime and ...'

11.1 FORMULATION

Essentially, only two John Innes composts were formulated; one with a low nutrient content was used for seed sowing and the rooting of cuttings; the other was a potting compost for general pot work. To allow for the varying degree of plant vigour and for seasonal changes in the growth rates, the nitrogen, phosphorus and potassium levels in the potting compost were adjusted by varying the rate of application of a 'complete' base fertiliser containing N, P and K. The formulae of the composts are as follows:

SEED COMPOST

Bulk ingredients (by volume)	Fertilisers	per cubic yard	per cubic metre
2-loam	Superphosphate	2 lb.	1·186 kg
1-Peat	Calcium carbonate	1 lb.	0·593 kg
1-Sand[1]			

POTTING COMPOST [JI Potting Compost No. 1 (JIP-1)]

7-loam	Hoof and Horn[2]	2 lb.	1·186 kg
3-Peat	Superphosphate[2]	2 lb.	1·186 kg
2-Sand[1]	Potassium sulphate[2]	1 lb.	0·593 kg
	Calcium carbonate	1 lb.	0·593 kg

[1] Although the term 'sand' is conventionally used, the particles grade up to $\frac{1}{8}$ in. diameter and the term 'grit' would be more correct.

[2] It is possible to purchase the hoof and horn, superphosphate and potassium sulphate as a complete ready-mixed fertiliser, it is then known as the 'John Innes Base' fertiliser.

Where higher nutrient levels are required, the rates of hoof and horn, superphosphate and potassium sulphate (or, for convenience, the JI Base fertiliser) are doubled to give the JI Potting Compost No. 2 (JIP-2), or trebled to give the JI Potting Compost No. 3 (JIP-3); the rates of calcium carbonate are also increased accordingly.

11.2 COMPOST INGREDIENTS: LOAM

Loam is the most important of the ingredients in the JI composts. Apart from its effect upon the physical and biological properties of the compost, it forms the basis of plant nutrition by supplying clay, which has a cation and anion exchange capacity, micro-elements and organic matter, which gives a slow release of nitrogen to the plants. Loam is also the most variable of the ingredients in the compost, both physically and chemically, and careful attention must therefore be given to its selection and preparation, including steam sterilisation, if good results are to be consistently obtained.

SELECTING A LOAM

The most important factors to consider when selecting a loam are: pH, texture, structure, organic content, and response to steam sterilisation.

pH. Loams chosen for making the John Innes composts should have a pH within the range 5·5 to 6·5. The pH of the loam is important for two reasons: its effect upon the mineral nutrition of plants, and its effect on the reaction of the loam to steam sterilisation. When loams have a pH below 5·5, there is the risk of a low level of fertility; the rate of phosphorus fixation may be high and there is also the possibility of a large release of manganese following steam sterilisation. If the loam has a pH above 6·5, there is the possibility of microelement deficiencies such as boron and iron and the very high risk of nitrogen toxicities developing after steam sterilisation.

Acid loams must be limed to bring the pH up to 6·3. If the actual amount of lime required to raise the pH to this value is not known, an estimate of the amount of calcium carbonate required can be made using the following formula:

desired increase in pH × 18 =
        ounces of calcium carbonate per sq.yd per 9 in. depth

e.g. for a loam with a pH of 5·6,

(6·3 − 5·6) × 18 = 0·7 × 18 =
        12·6 ounces calcium carbonate sq.yd per 9 in. depth

Whilst the actual lime requirement for a given change in pH will depend upon a number of factors, of which the clay and the organic matter contents are the most important, the above formula has been found to give reasonable estimates of the lime requirement of soils having about 20% clay content. Loams with a pH above 6·3 are better avoided and under no circumstances should loams having a pH of 7 or more with free carbonates present be used for compost making.

The effect of the pH on the biological and chemical changes induced by the heat sterilisation of the soil are not always fully appreciated. The risk of manganese toxicity occurring if the pH is low has to be balanced against ammonium toxicities occurring at high pH values; unless great care is exercised, it is possible to move away from one form of toxicity into another. The importance of adequate pH control of the soil cannot be overemphasised. Manganese toxicity is discussed more fully in Chapter 13 and some of the interactions which occur between steaming, liming and organic forms of nitrogen are described on page 220.

Texture. This term refers to the composition of the soil, i.e. the amounts of clay, silt and sand as determined by a mechanical analysis. A loam suitable for compost making has about 20% clay particles and is known as a medium-clay loam. The clay content not only helps in giving soils a good physical structure, it also forms the basis of plant nutrition by providing a cation and anion exchange capacity. A medium-clay loam can be identified by its ability, when moist, to take on a smear or polished surface when rubbed between the fingers. Sand gives a gritty or rough feel, whilst silt has a smooth or silky feel but does not give a polished surface.

Structure. This refers to the grouping of the particles of clay, silt, sand and organic matter into small aggregates. It is also desirable that these aggregates have a good stability and do not breakdown and disperse when the soil is wet. It has long been known that grass roots improve the soil structure; a permanent ley has a much better structure and aggregate stability than an arable soil of similar texture. Alvey (1961) studied the way in which growing four grass species affected the aggregate stability of three arable soils having clay contents ranging from 18 to 29%. On each of the soils, separate plots were sown with either rye grass, cocksfoot, fescue or timothy, and an additional plot was maintained in a fallow condition. Each treatment was assessed annually over a four year period for its effect on the mechanical stability of the soil aggregates which was measured by a wet sieving technique. With all three soils, the grass treatments showed an increase in the aggregate stability by comparison with the plots maintained in a fallow condition. When the soils were used to make John Innes composts, the grass treatments also showed a greater release of mineral nitrogen; this was derived from their higher organic matter contents.

Organic matter. In addition to its indirect effect in helping to improve soil structure, organic matter also provides a supply of plant nutrients, principally of nitrogen. The organic matter should be present partly as fibrous, relatively undecomposed grass roots and partly in the more decomposed form as humus. A loam with a good organic matter content will give a steady release of nitrogen over a long period. For this reason leys are preferable to arable soils for compost making.

Response to steam sterilisation. Steam sterilisation of the loam to eliminate the fungi and insects, which would otherwise attack the seedlings, is an essential part of the preparation of the John Innes composts. Unfortunately, in addition to eliminating the pathogens, the steam sterilisation of some soils can lead to the production of substances harmful to plant growth. These toxins can be of mineral or organic origin and, whilst the risk of toxicity can be minimised by avoiding loams with either a high or low pH and those with a high organic matter content, the effect of steam sterilisation on seedling growth cannot be precisely predicted from either a visual examination of the loam or its chemical analysis. Whenever a new loam is to be used for compost making, it is prudent to first make a small growing test, using seedlings, such as antirrhinum, to compare growth in the steam sterilised new loam with similar seedlings grown in another loam whose reaction to sterilising is already known. If growth in the new loam is comparable to that in the other soil and no toxicities are apparent, it is then safe to proceed.

MAKING A LOAM STACK

Having selected a loam which meets the above requirements, it should then be made into a loam stack; this presents an opportunity of adjusting its pH level and increasing its general fertility. Turfs are cut to a depth of 4 in. in the spring of the year when the loam is still moist and the growth is active. A loam stack about 6 ft square is made by first forming a layer of loosely packed turfs, grass-side facing downwards; upon this is placed a 2 in. layer of strawy animal manure, which is followed by another layer of turfs. A previously made chemical analysis is used to decide whether any phosphorus or potassium should be added; if the analysis shows these elements to be very low, 1 oz. each of superphosphate and sulphate of potash per square yard is added to each layer of the loam stack. If the pH is below 6·3, lime is added according to the formula given on page 217, the lime and the strawy animal manure being added alternatively between the layers of turfs. If the loam is in a dry condition when the heap is being made, it should be watered generously, otherwise the animal manure and the grass roots will not decompose sufficiently. When the loam stack cannot be built under cover, some protective covering must be erected in time for the heap to dry out before it is sterilised; wet soils cannot be sterilised successfully. The loam will be ready for use in about six months after stacking. It is prepared for sterilising by first chopping it down in thin layers to minimise any variation within the heap. After this it is put through a mechanical shredder or a hand sieve.

11.3 PEAT

The function of peat in the John Innes composts is primarily that of a physical conditioner. It increases the total porosity and improves both the water retaining capacity and the aeration. Either sphagnum or sedge peat having particles grading from $\frac{1}{8}$ in. to $\frac{3}{8}$ in. can be used, but fine dusty peats and so-called 'rhododendron' peats with a low pH are not suitable. The pH of the peat should be between 3·5 and 5·0, and unlike loamless composts where the peat usually forms the greater part of the compost's bulk, additional lime to correct for the acidity of the peat is not required; liming the loam to pH 6·3 is sufficient. The small quantity of calcium carbonate included with the base fertiliser serves only to neutralise the acidity developed by the fertilisers. The peat is not normally sterilised but it must be moistened before being used; air-dry peat does not mix well with the loam and the sand.

11.4 SAND

Sand is used as a physical conditioner. To enable it to perform its function of allowing excess water to drain from the compost, it must have a coarse grading with about 60–70% of the particles being between $\frac{1}{16}$ and $\frac{1}{8}$ in. diameter.

Technically, it would be more correct to use the term 'grit' for material of this size, but the term 'sand' has traditionally been used with reference to the John Innes composts; the true horticultural sands with fine particle sizes are not suitable for use in the John Innes composts. Apart from its particle size, it is also important to ensure that the sand is free from carbonates; the use of sand containing carbonates can result in a high pH with all its associated nutritional problems.

11.5 STERILISATION

To make a John Innes compost *it is essential that the loam is sterilised,* and providing it is done efficiently, heat sterilisation, by means of steam or electricity, has given better results than sterilisation with chemicals. Two types of toxicity can occur as a result of sterilising the soil; manganese toxicity is associated with some types of soils having a low pH, whilst nitrogen toxicity may occur when soils rich in organic matter and having a high pH are sterilised. Sterilisation speeds up the biological breakdown of organic matter and also introduces a temporary blockage in the nitrogen cycle. Those micro-organisms in the soil which convert ammonium to nitrites and then into nitrates are eliminated by steaming and, for a period, their re-establishment is inhibited, thereby resulting in a build-up of ammonium which may reach toxic levels.

An example of the interactions which can occur between heat sterilisation, the soil pH and the amount of readily nitrifiable organic matter is shown in Figure 11.1. Tomato seedlings were pricked-out into loam-based John Innes type composts; the unlimed soil had a pH of 5·17 and from previous experience it was known to be of a type which did not produce manganese toxicity after it had been steamed. Composts were made in which the soil was: (a) unlimed and not sterilised, (b) limed and not sterilised, (c) unlimed and sterilised, and (d) limed and sterilised. Nitrogen in the base fertiliser was supplied from either of two sources, (1) hoof and horn or (2) urea-formaldehyde. The results show that, when the nitrogenous fertiliser was hoof and horn, neither steaming nor liming by themselves caused any significant degree of toxicity but when the soil was limed to pH 6·3, steam-sterilised and then made into a compost, nitrogen toxicity occurred. This toxicity was proportional to the amount of readily mineralised nitrogen present (Fig. 11.2). When the nitrogen source was urea-formaldehyde, however, no toxicity developed; this was because of the slower rate of nitrogen availability. Other experiments have shown that toxicity can occur even when the amount of hoof and horn fertiliser added to the compost is low but the amount of organic matter in the loam is high, i.e. toxicities result whenever the level of readily nitrifiable organic matter is high, irrespective of whether it originates from a nitrogenous fertiliser or from using soils such as old cucumber beds which are high in organic matter.

Fig. 11.1 The effects of steam sterilising and liming the loam, when the compost contains a large amount of organic nitrogen. *Top left:* not steamed, not limed. *Top right:* not steamed, limed. *Bottom left:* steamed, not limed. *Bottom right:* steamed, limed.

From other experimental results obtained by the author (Bunt, 1956), it was found that the breakdown of organic matter and the release of ammonium is not affected by the sequence of adding the lime, i.e. the same results are obtained irrespective of whether the lime is added immediately before or after sterilisation.

MAKING THE COMPOSTS

A stratified heap of shallow layers of the measured ingredients is made on a clean surface, a small quantity of sand being retained to be first mixed with the fertilisers and so act as a carrier. This mixture is then spread uniformly over the heap which is turned by hand four times; alternatively a rotary mixer can be used.

11.6 CHARACTERISTICS AND USE

The concentration of the principal plant nutrients in the compost, i.e. the N, P and K, can be adjusted by increasing the quantity of base fertiliser, and the choice of compost strength is determined by the vigour of the crop and the season. For example, primula seed is sown in JI-Seed compost and pricked-

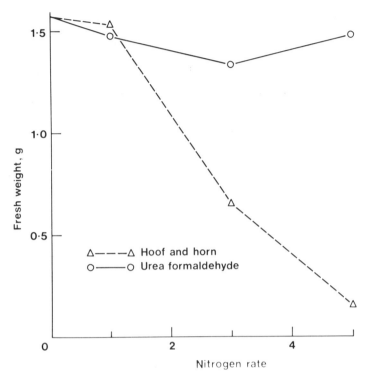

Fig. 11.2 The effect of nitrogen source and rate on the growth of tomatoes in a loam compost. When a readily mineralised source of nitrogen such as hoof and horn was used, the depression in growth was related to the amount of nitrogen. With urea-formaldehyde there was no depression in growth even at the highest rate of nitrogen. The loam had been steamed and limed for both sets of composts.

out into JIP-1 compost, whereas cucumber seed is germinated in JIP-1 and potted into JIP-2. Tomatoes are pricked-out into JIP-1 compost in winter and into JIP-2 compost in spring, and for plants such as tomato and chrysanthemum which are grown to the fruiting or flowering stage in 10 in. pots, the JIP-3 compost is used.

NUTRIENT LEVELS

The effects of increasing the strength of the base fertiliser on plant growth was studied by Bunt (1963). Tomatoes were grown in the three potting composts and also in sand cultures which received a nutrient solution at each watering. The plants were cut and weighed at regular intervals and the tissue analysed for the total N, P and K present. Under the environmental conditions prevailing in this particular experiment it was found that for all strengths of JI composts the maximum increment of growth occurred in the period 30 to 37 days after pricking-out (Fig. 11.3). Increasing the strength of the fertiliser did not prolong the period of high growth rate; irrespective of the fertiliser level the growth rate declined at the same time. By contrast, the plants growing in the sand cultures and receiving nutrient solution at each irrigation continued to show an increase in their rate of growth throughout the experiment. Analysis of the

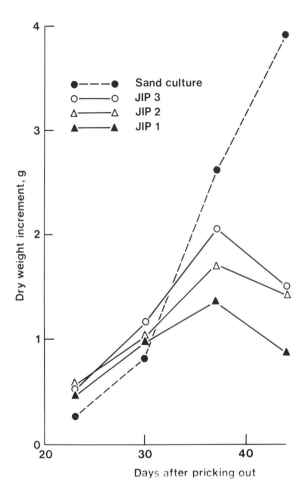

Fig. 11.3 The effect of increasing the amount of base fertiliser in the John Innes composts on the growth of tomatoes, in comparison with plants grown in sand cultures and given frequent applications of nutrient solutions. Increasing the amount of base fertiliser did not prolong the period of rapid growth; the maximum growth increment occurred in the same period for all fertiliser strengths. With the sand cultures, the growth rate was maintained by frequent liquid feeding.

plant tissue (Fig. 11.4) showed that with all the JI composts nitrogen was the limiting factor; increasing the amounts of base fertiliser in the compost did not materially affect the rate at which the nitrogen percentage in the plant tissue decreased with time. Only in the sand cultures where nutrients were being supplied continually was the growth rate and nitrogen level in the plant tissue maintained. It was concluded that the principal effect of increasing the

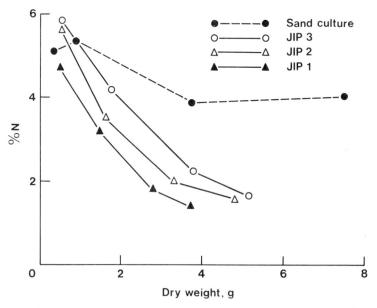

Fig. 11.4 Tissue analysis of the treatments in Fig. 11.3 show that increasing the amount of base fertiliser did not maintain the nitrogen level, whereas continual liquid feeding did.

rate of base fertiliser was to increase the potential *rate of growth* rather than to prolong *the period of growth* before feeding was required. Where the limiting factor is the length of the period over which the nitrogen in the compost is mineralised, this can only be extended by using a slow-release form of nitrogen, e.g. urea-formaldehyde, or by applying liquid fertilisers; it cannot be extended by increasing the quantity of hoof and horn.

SEASONAL EFFECTS

The importance of relating the strength of the compost to the prevailing light intensity has been studied by growing plants at different light intensities in controlled environment rooms (Bunt, 1963). Three levels of light intensity were used, i.e. 300, 600 and 900 lumens per sq. ft, and the rate of growth was determined from fresh weight measurements made at fourteen, twenty-eight and forty-two days after pricking-out (Table 11.1). At the lowest light intensity, plants grown in the JIP-3 compost made significantly less growth throughout the experiment than those in the JIP-1 compost. At the highest light intensity, i.e. 900 lumens per sq. ft, growth at fourteen days after pricking-out was suppressed by the higher level of fertilisers; at twenty-eight days after pricking-out there was no significant difference in growth between the plants

Table 11.1. *The interaction of light intensity and compost nutrient levels on the growth of tomato (fresh weight, grams).*

Light Intensity (lumens/ft^2)	Harvest 1, 14 days			Harvest 2, 28 days			Harvest 3, 42 days		
	JIP-1	JIP-2	JIP-3	JIP-1	JIP-2	JIP-3	JIP-1	JIP-2	JIP-3
300	0·366	0·281	0·172	2·42	1·93	1·54	13·14	14·30	10·27
600	0·696	0·644	0·449	6·92	6·71	6·21	21·93	28·38	30·31
900	0·915	0·892	0·579	8·14	8·44	8·81	22·93	29·28	34·36

grown in the two composts, whilst at forty-two days after pricking-out plants in the JIP-3 compost had achieved about 50% more growth than those in the JIP-1 compost. It can be seen from this experiment that there will be no advantage in using composts with a high fertiliser strength during the low light conditions of winter. Under these conditions, an increase in the base fertiliser rate will actually depress plant growth; better results will be obtained by using a compost with a low rate of base fertiliser and then supplementing this with liquid feeding as the plants increase in size and the light intensity improves.

STORAGE

Mineralisation of the nitrogen present in the organic matter in the loam and in the hoof and horn fertiliser is a biological process, the rate of which is primarily controlled by the moisture content, the pH and the temperature of the compost. High levels of moisture and temperature will significantly increase the mineralisation rate. If the composts are stored after mixing, it means that the pH will first increase because of the production of ammonium and then decrease as nitrates are formed (Bunt, 1956). During this process, the pH of a JIP-2 compost can rise by 0·7 of a unit above its initial value and then, as the ammonium is converted into nitrates, fall to about 1·0 unit below the initial value. This means that pH determinations of the compost made after mixing are meaningless unless the exact stage of nitrogen mineralisation is known.

Also, as the nitrogen is converted from an organic form to inorganic forms, the osmotic concentration is increased. The build-up in the ammonium concentration is not, however, revealed by the conventional conductivity determination; only as the nitrates are formed does the specific conductivity increase. Typical salinity values found for freshly mixed and stored John Innes composts are:

Compost	EC_e millimhos/cm at 25°C	
	Freshly mixed	After storage and nitrate formation
JIP-1	3·69	5·36
JIP-2	4·56	6·84
JIP-3	5·58	8·10

With all composts, the salinity increases by approximately 50% during storage due to the nitrification of the organic nitrogen. This increase in salinity can be detrimental to seedling growth. Usually seedlings pricked-out into stored composts will show a check in their growth by comparison with those grown in freshly mixed composts.

Some measure of the effect of storage of the compost and the stage of nitrogen mineralisation on the growth of young seedlings can be seen from the results of an experiment where tomato and antirrhinum were grown for a short period in freshly prepared composts, the nitrogen source being either: urea-formaldehyde, or hoof and horn, or ammonium sulphate or calcium nitrate (Fig. 11.5). With urea-formaldehyde, there was no reduction in the growth of

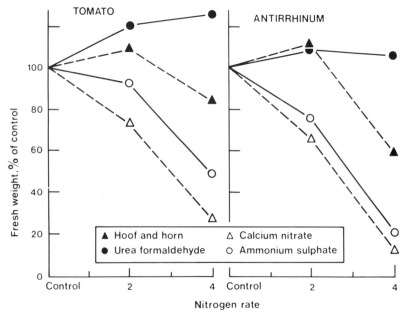

Fig. 11.5 Comparative effects of increasing amounts of different sources of nitrogen on the growth of tomato and antirrhinum.

either species as the amount of nitrogen was increased; this was because of its slow rate of nitrogen mineralisation. With hoof and horn, the more rapid rate of ammonium production caused some reduction in growth as the amount of nitrogen was increased whilst, with ammonium sulphate and calcium nitrate, an increase in the amount of nitrogen gave a substantial reduction in the amount of growth. With the two latter fertilisers, it was found that their adverse effect on plant growth was directly related to their effect on increasing the salinity of the compost. At an EC_e of approximately 5 millimhos, tomatoes made over 3 g of fresh weight per plant; as the salinity of the compost was increased by the higher rates of fertilisers, growth decreased progressively until, at an EC_e of 9·5 millimhos, the mean fresh weight was less than 1 g per plant. Results in Figure 11.5 have been expressed as a percentage of the weight of seedlings grown in the control compost with no nitrogen added. This method of presentation enables the response of the two species to the various forms of nitrogen to be compared. Antirrhinum was found to be more sensitive to both the rate and form of nitrogen than the tomato.

It will be seen from the various experiments that, whenever possible, the composts should be used without being stored, and *in no circumstances should they be stored for more than three weeks.*

11.7 COMPOSTS FOR CALCIFUGE PLANTS (JIS(A))

It is well known that calcifuge plants, such as ericas and rhododendrons, do not thrive in composts which have a pH approaching the neutral point; often the plants turn chlorotic and make little growth. This problem was investigated by Alvey (1955) in relation to the John Innes composts. *Erica gracilis* was grown in both the standard John Innes seed compost and the potting compost (JIP-1). These composts were also modified by replacing the $\frac{3}{4}$ oz./bushel of calcium carbonate, normally added with the base fertiliser, with flowers of sulphur at either $\frac{3}{4}$ or $1\frac{1}{2}$ oz./bushel. Other treatments included a ferrous sulphate solution at 1 oz./gall. watered on to the compost, and foliar sprays of ferrous and manganese sulphates. Whilst the amount of growth in the standard JI-Seed and JIP-1 composts was good by comparison with a typical erica compost, the incidence of chlorosis was high, especially in the JIP-1 compost. In composts where the calcium carbonate had been replaced with sulphur, there was good growth and chlorosis was virtually eliminated. The fresh weights, chlorosis rating, and pH of the composts are given in Table 11.2. At the highest rate of sulphur, there was a depression in growth; for this reason, the seed compost with the lower rate of sulphur was chosen as being the best compost for growing ericaceous plants and was described as JIS(A), the (A) denoting the non-standard, acid compost.

Bunt (1956) examined the rate at which sulphur was oxidized in composts and found that, under typical glasshouse conditions of high soil temperatures

Table 11.2. *Growth of* Erica gracilis *in different composts.*

Compost	Fresh weight (grams)	Chlorosis rating	pH At Potting	pH After Cutting
JIS	1 490	53	6·4	6·6
JIS + ¾ oz. sulphur	1 604	8	5·7	4·7
JIS + 1½ oz. sulphur	751	0	5·7	3·9
JIP-1	1 021	139	6·5	6·1
JIP-1 + ¾ oz. sulphur	1 719	53	5·7	5·0
JIP-1 + 1½ oz. sulphur	1 483	0	5·8	4·2
'Grower's'	907	30	4·4	4·9

and moisture levels, the maximum pH reduction in the compost occurred within 6–8 weeks.

Chapter 12

Heat sterilisation

Practically all of the soil-borne pests and diseases which attack seedlings and pot-grown plants can be eradicated by heat treatment of the materials before they are used for making the composts. This method of controlling soil pathogens was first used by glasshouse growers at the end of the nineteenth century and rapidly increased in importance until the 1950s, when virtually all the mineral soil used for making potting composts was heat sterilised. Since this time, the increasing use of loamless composts made from materials such as peat, perlite, vermiculite, plastics, etc., which are virtually free of pathogens and therefore do not require sterilising, has meant that heat sterilisation is not such an important factor in the preparation of pot plant composts as it once was.

12.1 THERMAL DEATHPOINTS

The temperature required to kill an organism is known as its thermal deathpoint. Usually a precise temperature cannot be stated, as this largely depends upon three factors: the form in which the pathogen is present (resting bodies and eggs are slightly more resistant to heat than the active stages); whether the heat being applied is moist or dry (moist heat is more effective than dry heat); and, the most important factor of all, the thermal deathpoint is closely related to the duration of the heating period. Commonly accepted values of thermal deathpoints are 120°F (49°C) maintained for 10 minutes for nematodes and eelworms, and 122°F (50°C) for most of the fungi, although some wilt fungi require 180°F (82°C) for 10 minutes. Insects and slugs are killed at 150°F (66°C) and most weed seeds are killed at 170°F (77°C) maintained for 10 minutes; a few seeds, however, are known to survive at 212°F (100°C). Whilst tomato mosaic virus (TMV) present in root tissue is inactivated at 185°F (85°C) for 10 minutes, in dry debris it can survive a dry heat of 212°F (100°C) for 20 minutes. Sheard (1940) determined the thermal deathpoints of a number of fungi and eelworms under both laboratory and practical nursery conditions. He found that whereas pathogens growing in pure cultures in the laboratory were killed at 140°F (60°C) for 10 minutes, under typical nursery conditions,

Heat Sterlisation

a temperature of 158°F (70 C) for 10 minutes was required to obtain a complete kill. This apparent increase in the thermal deathpoint under practical conditions was attributed to the difficulty of ensuring that *all* the soil had reached a temperature of 140°F (60°C); small pockets of soil which were slow to heat up allowed some survival of the pathogens. For this reason the higher temperature, which offers a safety margin, was recommended for nursery conditions.

The relationship of the thermal deathpoint to the duration of the heating period is of considerable practical significance to growers. By heating the soil to a lower temperature for a longer period, some of the chemical and biological reactions which occur at high soil temperatures, and which can be detrimental to plant growth, are thereby avoided or reduced. This relationship is shown in Figure 12.1 for the chrysanthemum eelworm (*Aphelenchoides*

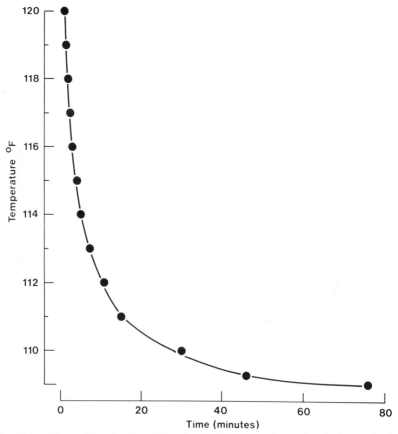

Fig. 12.1 Thermal deathpoint of the chrysanthemum eelworm in relation to the time the temperature is maintained.

ritzema-bosi). Whilst the actual values given are specific for this particular organism, the curve is typical of that found for a number of pathogens and shows how a long heating period at a low temperature can be used in place of high temperatures for shorter periods. The temperature × time relationship can be expressed mathematically as an inverse square root curve:

$$(Y - c)\sqrt{X} = k$$

where Y = the lethal temperature, X = the duration of the heating period, and c and k are constants.

All of the organisms present in soils are not killed until a temperature of 257°F (125°C) is reached and, with a controlled form of heating such as steam, the temperature does not exceed 212°F. It would therefore be more correct to use either of the terms 'partial sterilisation' or 'pasteurisation' to describe the heat treatment; the term sterilisation is, however, widely accepted and used by growers, and has been retained for that reason.

12.2 METHODS OF HEAT STERILISATION

The most commonly used methods of heat sterilisation are: steam, steam-air mixtures, flame pasteuriser, and electricity. In the past, two other methods have also been used, viz. baking, and boiling water, but because of harmful side effects neither of these methods is recommended. Where a steam supply is available, e.g. from a glasshouse heating system, this is an obvious choice of method, especially if large quantities of sterilised soil are required. When there is no steam supply or when only small quantities of sterilised soil are required, flame pasteurisers and electric sterilisers are convenient alternatives.

12.3 STEAM

The principal advantages of using steam for soil sterilisation are, (a) it is a concentrated form of heat and is easily conveyed from the boiler to the sterilising area, and (b) it gives control over the final soil temperature. With this form of heating, the temperature cannot rise above 212°F, and providing that the soil has been correctly prepared and the steam is 'dry', the treated soil can be used for making composts immediately the heating process has finished and the soil has cooled.

STEAM GENERATION

When 1 unit of heat, i.e. a British Thermal Unit [BTU (1 BTU = 1·055 kJ)], is applied to 1 pound of water the temperature is increased by 1°F. At atmospheric pressure, water boils at 212°F and boiling water contains 212 − 32 = 180 BTUs of sensible heat. If more heat is applied to the water, there will be no further increase in the temperature but the water will be converted into steam.

Each pound of steam at atmospheric pressure contains 970 BTUs of latent heat plus the 180 BTUs of sensible heat, making a total of 1 150 BTUs. It is the latent heat in the steam which is available for sterilising the soil. When steam is generated under pressure, i.e. in a boiler, the temperature at which the water boils is increased, thereby producing steam at a higher temperature and with a slightly increased amount of sensible heat; there is also a slight drop in the amount of latent heat. For example, water at a pressure of 60 lb./sq. in. boils at a temperature of 307°F, and 1 lb. of steam at this pressure contains 277 BTUs of sensible heat and 905 BTUs of latent heat, giving a total heat content of 1 182 BTUs. The space occupied by 1 lb. of steam also decreases from 26·8 cu.ft at atmospheric pressure to 5·8 cu.ft at 60 lb./sq.in. The principal advantage to the grower of using high pressure steam is the decrease in the size of the steam pipe required to link the boiler to the sterilising area. When the steam is injected into the soil, it returns to practically atmospheric pressure and hence the soil temperature must be 212°F; there will be no difference in soil temperature resulting from the pressure at which the steam is used.

A secondary advantage of generating high pressure steam is the drying effect or the reduction in the amount of free water present in the steam which occurs as the pressure is reduced; this is technically known as the 'wire-drawing' effect. Steam generated in the manner described is not superheated and is known as 'saturated steam'. if it does not contain any free water droplets it is known as 'dry saturated steam'. In practice, however, boilers do not produce absolutely dry steam; also there is inevitably some loss of heat with condensation occurring in the steam mains. Depending upon the type of boiler, the length of the steam main and the thermal efficiency of the insulation, the steam which is used to sterilise the soil may contain up to 10% by weight of water. The additional heat present in the high pressure steam is used up by drying the steam or evaporating some of this free water as the steam pressure is reduced from 60 lb./sq.in. to atmospheric or zero pressure. In practice, it is very important that the steam used for sterilising is dry. Wet steam can create puddled or waterlogged patches in the soil which result in either inefficient sterilisation due to lack of steam penetration, or soil which is too wet to handle. To ensure that the steam is dry, a steam separator should be used to remove any free water (p. 234).

HEAT REQUIREMENT OF SOILS

The principal factors determining the amount of steam required to sterilise a given volume of soil are its heat capacity, and the thermal efficiency of the heating process.

The heat capacity of a soil comprises the dry matter, which has a specific heat of 0·2, and the soil moisture, which has a specific heat of 1·0; thus to heat an oven-dry soil would require only one fifth of the amount of heat required to heat an equal weight of water. In practice, it is found that soils which are very dry (dust dry) can be difficult to steam because of the slow rate at which the steam is condensed. Wet soils, having a high heat requirement, are also difficult to steam satisfactorily because of the restricted spread of steam; 'cold spots' and puddling of the soil develop leading to the blowing of uncondensed steam through the surface. In practice wet soils usually create more problems than dry soils. A light sandy soil in a suitable condition for steaming, having 12% by weight of water, has a heat requirement of 26 BTUs/cu.ft/°F (1 744 kJ/cu.m/°C), whilst a heavy clay soil, with 28% moisture, has a heat requirement of 32 BTUs/cu.ft/°F (Morris, 1954). The theoretical steam requirement per cubic foot of soil will therefore be the product of the temperature rise times the heat capacity. For example, a heavy clay soil, at a temperature of 50°F and having a heat capacity of 32 BTUs/cu.ft, has the following calculated steam requirement:

$$162 \times 32 = 5\,184 \text{ BTUs} \div 970$$
$$= 5\cdot3 \text{ lb. steam};$$

and for a light sandy soil:

$$162 \times 26 = 4\,219 \div 970$$
$$= 4\cdot3 \text{ lb. steam.}$$

For practical purposes we may say that the heat requirement of the soil increases by 0·75 BTU/cu.ft/°F for each 1% increase in its moisture content, and an average theoretical heat requirement for potting soils would be 5 lb. of steam per cu.ft of soil.

THERMAL EFFICIENCY

The thermal efficiency of the heating process depends upon:

(a) the physical state and moisture content of the soil,
(b) the quality or dryness of the steam,
(c) the design of the equipment.

When steam is introduced into the soil, usually by means of a perforated pipe or grid buried in the soil, it immediately condenses on to the cold soil adjacent to the steam orifice and releases its latent heat, thereby raising the soil temperature to 212°F. As more steam is injected into the soil, it travels through the pore spaces in the already heated soil until it reaches cold soil where it condenses. In this way a very narrow heat front, only about 1 in. thick, spreads through the soil. Irrespective of the steam pressure in the steaming grid, the pressure of the steam in the soil will not exceed 0·5 lb./sq.in. and it will often be only half this value. For practical purposes, the steam can be considered as being at atmospheric pressure and hence at a temperature of 212°F. It therefore follows that, when steam is in contact with soil, the temperature of the heated soil must be 212°F (100°C). Whenever a temperature of 180°F is quoted in connection with steam sterilisation, this can apply only to very small areas where, because of a poor physical condition, steam penetration has not occurred and heating has been by conduction. Under good sterilising conditions, heating by conduction is minimal and the bulk of the soil will be at 212°F. If the free pore spaces in the soil are reduced either by compaction, lack of aggregate stability or by water, and the steam is unable to spread freely, the steam pressure in the soil builds up to the point where it blows up through the surface and escapes; for this reason a dry soil in good tilth is necessary for efficient sterilisation. Localised puddling of the soil and steam loss by 'blowing' can be caused by free water introduced into the soil by 'wet' steam. To ensure that the steam is 'dry', a steam separator must be installed at the end of the steam line. The separator operates on the principle of creating a rotary flow of the steam, which allows any water present to be separated by centrifugal force; this falls to the bottom of the separator, where it is released by a thermostatically controlled steam trap.

Assuming the above requirements regarding soil condition and steam quality have been met, the amount of steam required under practical conditions will range from 5·5 to 7 lb./cu.ft. of soil (Bunt, 1954, *a*), i.e. the thermal efficiencies will be of the order of 90 to 70%. Long heating periods give lower thermal efficiencies because of the loss of steam from the sides of the sterilising bin. Usually, rapid heating of the soil is desirable and on the basis of a ten-minute heating period and a steam requirement of 6·5 lb./cu.ft of soil, 1 cu.yd of soil will require 175 lb. of steam, with a steam flow rate of 1 050 lb. of steam per hour. Although 6·5 lb. of steam is required to heat each cubic foot of soil, the increase in the water content as a result of steaming will be equivalent to only approximately half the amount of steam used. This is partly because of some waste of steam during the heating process and partly because of the continued loss of water vapour from the soil after the steam has been turned off.

TREATMENT OF BOILER FEED WATER

To prevent the corrosion of the steam mains and condensate return pipes, which is caused by oxygen and carbon dioxide dissolved in the water, it is customary to treat the boiler feed water with corrosion inhibitors, and two types of amines are commonly used for this purpose:

Filming amine (octadecylamine) is an inert chemical which volatilises with the steam and condenses on the walls of the pipes, forming a film on the inner surfaces. Tests have shown that this material is not toxic to plants and it may be used safely as a corrosion inhibitor.

Neutralising amine (cyclohexylamine) volatilises and combines with the CO_2 in the steam. This material is toxic to plants and it can be used only in closed circuit heating systems, where no steam is ever used to sterilise soils. On glasshouse nurseries, it is always probable that steam will be wanted on some occasion to sterilise soils and this material is not therefore recommeneded as a corrosion inhibitor for glasshouse boilers.

12.4 STEAM–AIR MIXTURES

There are several reasons why it is desirable when heat treating soil that a temperature of less than 212°F should be used, the most important being:

(1) biological changes are reduced at the lower temperatures,
(2) changes in the organic and mineral form of plant nutrients following heating are reduced,
(3) less heat is required and therefore cost is lower.

It has already been shown that the thermal deathpoint of soil-borne pathogens can be safely reduced below 212°F by extending the length of the heating period. Biologists have long been aware that the natural biological control of diseases plays an important role, i.e. one organism is antagonistic to, and controls the numbers of another. When soils are heated to 212°F, both the pathogens and their natural antagonists are eliminated and if pathogens reinfect a recently sterilised soil they spread at a much higher rate because of lack of competition and cause a greater loss of plants than if the soil had not been treated. Baker (1967) has described the situation occurring in recently sterilised soils as being a 'biological vacuum'. He contrasts the loss of seedlings grown in a soil which had been steam-sterilised and then inoculated with the damping-off fungus *Rhizoctonia solani*, with another treatment where *Myrothecium verrucaria* was introduced at the same time as *R. solani*. In the first treatment, the pathogen destroyed most of the seedlings whereas, in the latter treatment, very few of the seedlings were killed because of antagonism and the containment of the pathogen by the other fungus.

One way of countering the problem of reinfestation is to reduce the temperature to which the soil is heated and so allow the natural antagonists to survive the heat treatment. It has already been shown that, when using steam as the heating medium, a temperature of 212°F in the bulk of the soil is inevitable; by introducing air into the steam supply, a steam–air mixture is created which has a lower temperature. On the occasion of the first recorded use of steam-air mixtures for the heat sterilisation of soil (Bunt, 1954, *b*), an industrial pattern steam injector operating on the venturi principle was used to entrain free air. The temperature of the mixture could be controlled by variable orifice steam- and air-nozzles. Since this time,

236 Heat Sterilisation

Baker and his co-workers have made detailed studies of both biological and engineering aspects of steam–air mixtures, and this method of soil sterilisation is now well established in the USA and Australia.

CALCULATION OF STEAM AND AIR REQUIREMENT

Some physical data by Aldrich *et al.* (1972) relating to steam-air mixtures of various temperatures, including the rates of steam and air flow necessary to treat one cubic yard of a 1:1:1, soil:peat:perlite compost within thirty minutes, are given in Table 12.1. The table has been constructed on the basis

Table 12.1. *Properties of steam–air mixtures.*

Mixture temperature ($°F$)	Steam–air ratio (lb. steam/ft^3 air)	Air–steam ratio (ft^3 air/lb. steam)	Heat available (BTU/ft^3 of air)
140	0·01124	89·0	12·9
150	0·01574	63·5	18·1
160	0·0225	44·5	25·7
170	0·0318	31·4	37·3
180	0·0487	20·5	56·8
190	0·0850	11·8	94·9

of a steam temperature of 230°F, i.e. at a pressure of about 5 lb/sq.in., and air at a temperature of 70°F, having a relative humidity of 50%; about 95% of the heat used in treating the soil will be obtained from condensing water vapour out of the steam. The required rates of steam and air flow for any soil temperature can be calculated from the data in the table.

For example, 1 cu.yd of compost, having a bulk density of 56 lb. weight cu.ft and a moisture content of 30%, is to be heated to 170°F in 30 minutes. The specific heat of the oven-dry compost is assumed to be 0.2.

Then: (a) 1 cu.yd will weigh $56 \times 27 = 1\,512$ lb.,
    (b) the weight of water (30%) will be 454 lb.,
    (c) the heat required to warm the compost to 170°F will be:
        $1\,512 \times 0·2 \times (170° - 70°) = 30\,240$ BTUs,
    (d) the heat required to warm the water will be:
        $454 \times 1·0 \times (170° - 70°) = 45\,400$ BTUs,
    (e) the total heat requirement will be:
        $30\,240 + 45\,400 = 75\,640$ BTUs,
    (f) the heat available from aerated steam at 170° is:
        37·3 BTU/cu.ft (Column 4),
    (g) the volume of aerated steam required is:
        $75\,640 \div 37·3 = 2\,028$ cu.ft,
    (h) assume a heating efficiency of 50%, then the rate of air flow required will be:

2 028 ÷ (30 × 0·50) = 135 cu.ft/min.,
(i) 1 cu.ft of aerated steam at 170°F contains 0·0318 lb. of steam (column 2),
(j) the steam flow required will be:
0·0318 × 135 = 4·3 lb./min.

The steam and air flow rates required to treat one cubic yard of compost in 30 minutes are given in Table 12.2.

Table 12.2. *Rates of steam and air flow required to treat 1 cu. yd of compost in 30 minutes (50% heating efficiency assumed).*

Mixture temperature (°F)	Steam (lb./min.)	Air (cu.ft/min.)
140	3·1	274
150	3·5	222
160	3·9	177
170	4·3	135
180	4·8	98
190	5·5	64
212	6·5	0

To avoid large cumbersome steam injectors, it is customary to use either air blowers or fans to supply the air for the mixture which should have a temperature of 150 to 160°F. It should be noted that, unlike sterilisation by pure steam *where rapid heating and cooling is desirable*, with the steam–air method, the temperature must be maintained for thirty minutes. The steam–air mixture can be injected into the soil either from a buried grid, which allows the heat-front to rise up through the soil, or under a plastic sheet held over the surface of the soil so that the heat-front travels downwards from the soil surface. The latter method is preferred as 'blow-out' problems caused by lack of lateral spread of the mixture are avoided.

Using this system of heat sterilisation, Baker (1967) has shown that the number of plant deaths following the reintroduction of *R. solani* into heated soil were significantly reduced when a temperature of 160°F had been used, whilst no losses occurred at all when the temperature used was 140°F This effect was attributed to the survival of the natural antagonists at the reduced soil temperatures.

Changes in the forms of nitrogen and amounts of exchangeable manganese produced by the method are less than those occurring from the use of pure steam; this is discussed more fully in a later section on the chemistry of sterilised soils.

238 *Heat Sterilisation*

12.5 FLAME PASTEURISER

Another method of heat sterilisation of soil is by the flame pasteuriser. This consists of a revolving steel cylinder mounted at a slight angle to the horizontal; the cylinder is open at both ends and rotates at about 50 rpm. Heat is supplied by a blow-torch, burning either paraffin or a light oil, mounted at the lower end, with the flame directed into the cylinder (Fig. 12.2). The mode

Fig. 12.2 Flame pasteuriser. Sterilised soil leaving the rotating drum (covered with a shield) passes through the riddle before being collected in the wheelbarrow.

of sterilisation is primarily by steam generated from the water present in the soil rather than by direct heat. Providing that the soil is sufficiently moist and that the equipment is used correctly, the soil temperature will not rise above 212°F. It can be demonstrated that no significant amount of burning of the soil occurs, by including small pieces of paper or dry straw with the moist soil. If the equipment is being operated correctly, the paper will not be charred or burnt.

In operation, moist loose soil is shovelled into the upper end of the cylinder and then falls against the hot sides as the cylinder revolves. Soil is in the cylinder for only 20–30 seconds and the temperature can be controlled by varying the size of the flame, the angle of the cylinder and the amount of water in the soil. If the soil is too wet, however, it will 'ball-up' and will not mix uniformly with the other ingredients. To achieve satisfactory results, the steriliser should be operated to give a soil temperature of 175°–190°F; a temperature of 160°F has not proved satisfactory because of the very short time the soil is in the steriliser. This method of sterilisation has the advantage of being a 'flow' rather than a 'batch' process, with consequently a high potential output. Sterilisers are available with outputs of up to 8 cu.yd/hr; the fuel requirement is about 1 gall/cu.yd of soil treated.

12.6 ELECTRICAL STERILISERS

Electricity can be used in two distinct ways to heat-sterilise soil, either:

(a) indirectly, by means of heating elements of the immersion heater type,
(b) directly, by passing an electric current between two electrodes buried in the soil.

IMMERSION HEATER TYPE

Heat can be generated by passing an electrical current through a wire having a high electrical resistance. If the heating element is mechanically protected and also electrically insulated, it can be used to heat-sterilise soil in much the same way as an immersion heater is used in a domestic kettle. The only essential difference is that whereas fluids, such as water, can circulate in relation to the heater and convection currents are produced, soils are not fluid and the heating is almost entirely by conduction. A small amount of heating does occur by steam generated from the moist soil which is in direct contact with the heaters.

The steriliser is usually in the form of a heat insulated box mounted on legs, the base being hinged and held in the closed position by quick-release clamps; this allows rapid removal of the soil after sterilisation. The heating elements consist of sheathed resistance wires embedded into large plates. In addition to giving mechanical protection to the elements, the plates also provide a large heating surface.

With this type of steriliser, heating is by conduction, and dry soils have a relatively low rate of heat conduction. When the particles of soil are joined by a film of water, the rate of conduction is appreciably increased but, as the heat capacity of the soil is also increased, a compromise in the soil water content must be reached. Kersten (1949), working with a silty-clay loam (27% clay), found that, at a moisture content of 12·3%, the thermal conductivity was 0·0020 calories second^{-1} cm^{-1} °C and the heat capacity was 0·436 calories cm^{-3} °C. When the moisture content was increased to 18%, the thermal conductivity was increased by 25% and the heat capacity by 18%. As the soil was made progressively wetter, however, the thermal conductivity did not rise as rapidly as the heat capacity. Soil to be sterilised by this method should not therefore be very wet.

The surface temperature of the heaters rises to about 572°F (300°C), resulting in a thin layer of overheated soil, whilst the bulk of the soil reaches only 158°F (70°C). In practice, this small volume of overheated soil does not appear to cause any harmful effects to seedling growth. The reason for this is not clear. Dilution of the small volume of overheated soil with the remainder of the soil which has not been overheated has been one suggestion; it also seems probable that, because of the very high temperatures attained in the soil, toxins are not produced at the same rate as when soils are heated to intermediate temperatures (p. 242). One advantage of this type of steriliser is that the electrical loading is constant throughout the heating period; also the method is quite safe because the soil is never electrically 'alive'.

Sterilisers of this type are available in two sizes, one which is suitable for commercial use has a capacity of ¼ cu.yd of soil and an electrical loading of 9 kW, a smaller version suitable for the amateur has a capacity of 1·3 cu.ft (1 bushel) and an electrical loading of 1·5 kW.

ELECTRODE STERILISERS

An alternative method of electrical soil sterilisation is to pass an electrical current between two electrodes buried in the soil; the water acts as an electrical conductor and a resistance heater. With this method of sterilisation, the soil is electrically 'alive' and adequate safety measures, such as an interlocking switch on the lid of the steriliser, must be employed to protect the user. Unlike the immersion heater type of steriliser, the electrical loading is not constant during the heating process. At the start there is a high electrical resistance and the current is low. As the heating progresses, the resistance drops and the current rises. With some designs of steriliser, the flow of current just before the heating is finished can be up to eight times as high as at the start; under some conditions of electrical supply this can be a considerable disadvantage. The electrical resistance of a particular soil is directly proportional to the distance apart of the electrodes and inversely proportional to their area. Normally, the electrodes are 10–12 in. apart. With some sterilisers it is possible to halve the current being used by switching alternate electrodes out of circuit after the soil has started to heat.

Another factor crucial to the success of the operation is the electrical conductivity of the soil. This is governed by the amount of moisture and salts it contains and also by its density. Dry, loose soil is a very poor electrical conductor, whereas a soil that is moist and contains some mineral salts is a good electrical conductor. During sterilisation, the soil can contract away from the electrodes, and firm packing when filling is necessary to maintain electrical contact. Soil to be sterilised by this method should therefore be moist, contain about 35% water and be firmly packed. If the soil has a low mineral salt content, a solution of magnesium sulphate (6 g/l or 1 oz.gall.), applied with a watering-can as the steriliser is being filled, will improve the conductivity and will not cause any harm to seedling growth. Heating ceases when the soil adjacent to the electrodes dries out, thereby breaking the flow of electricity; at this point, the temperature of the bulk of the soil will be about 170°F. Sterilisers of this type are available with a capacity of $1\frac{1}{4}$ bushels, operating on single phase 240 V a.c. system with a maximum loading of 25 amperes. A larger three-phase model with just over a $\frac{1}{2}$ cu.yd capacity ($12\frac{1}{2}$ bushels) has a maximum loading of 40 amperes per phase. With both the immersion heater and electrode type of sterilisers, the electrical consumption is just over 1 kW/cu.ft of soil, but this figure is very much dependent upon the design of the equipment and the condition of the soil.

In recent years, the increasing use of loamless composts which do not normally require sterilisation, together with the increased number of nurseries having a steam supply available, have resulted in a decline in the use of electrical soil sterilisers.

12.7 OTHER METHODS

Two other methods of heat sterilisation of soil can be mentioned for historical interest: they are baking, and boiling water.

Baking was widely used in the early days when steam and electrical equipment were not generally available. Moist soil was placed on a steel plate over a fire; at the commencement of heating, the soil was in fact being steam sterilised by steam generated from the water within the soil. During the later stages of heating, when the water had been evaporated, there was no control over the soil temperature, and often very high temperatures occurred. In some soils, marked toxicities to seedling growth resulted and it was necessary after sterilisation to leave the soil to 'recover' before it could safely be used.

Boiling water has also been used for heat sterilising soils, but its low heat content and the puddled state of the soil after treatment did not commend its use.

12.8 CHEMISTRY OF HEAT STERILISATION

It has long been known that substances toxic to plants can be produced when soils are heated. The extent to which seed germination and seedling growth is retarded in sterilised soil can be very variable. In some cases it is so small that it passes unnoticed, especially as the sterilising check to plants is usually only temporary and this is quickly followed by a period of faster growth, which soon produces larger plants than those grown in unsterilised soil. In other cases, the check can be so great that it leaves a lasting effect on the growth of such plants as alyssum and antirrhinum; in extreme cases it may even result in their death.

It was mistakenly thought at one time that only one specific toxin was produced when soils are heated and attempts at its identification lead to apparent contradictory results. Whilst the full extent of the chemical and biological changes following heat sterilisation are still not known, it is possible to group the probable causes of toxicity and to suggest remedial action.

NITROGENOUS COMPOUNDS

Early workers (Pickering, 1908; Johnson, 1919) attributed the toxicity to an increase in the soluble organic matter and ammonia production. When soils are heated to 100°C, the nitrifying bacteria are killed and even if re-inoculation is made immediately after sterilisation, they do not establish and multiply for some time. The spore forming ammonifiers, however, are not suppressed and following heat sterilising there is usually a build up of ammonium before nitrates are once again formed in significant amounts. Johnson examined the effect of different temperatures on the production of toxic substances. In soil heated to 50°C, there was no delay or inhibition in the germination of lettuce seed by comparison with seed sown in the untreated soil. Between 100°C and 250°C, however, there was a considerable delay in germination, but when the sterilising temperature was increased to 350°C, germination improved again and showed only a slight delay by comparison with the control treatment. The effect of the temperature of the heat treatment on seed germination was closely related to the amount of ammonium produced; this rose from 2·5 mg per 100 g in the unheated soil to 13·9 mg in the soil heated to 250°C, and then declined to 7·3 mg in the soil heated to 350°C. He was also able to show that when working at a temperature of 110°C plant growth (weight) decreased as the heating period extended beyond ten minutes. There was not always consistency between results obtained from different soils, however; *tomatoes grown in peat heated to 110°C showed a decrease in growth of 200%* by comparison with plants grown in unheated peat, whilst *with a silt loam, heating gave a 254% increase in growth*. It was suggested that ammonia was the toxic agent. This work was extended by Lawrence at the John Innes Horticultural Institution in the 1930s, and resulted in a better understanding of the practical requirements for successful soil sterilisation.

Recent work has shown conclusively that high concentrations of ammonium (Maynard and Barker, 1969; Nelson and Hsiek, 1971) and of ammonia (Bennett and Adams, 1970) in soils can be toxic to plants. Below pH 6·0, very little free ammonia will be present in the compost; it will mostly be present in the ammonium (NH_4^+) form. As a result of heat sterilisation and the temporary blockage in the nitrogen cycle, ammonium produced by the biological decomposition of the natural organic matter present in the soil and from such nitrogenous fertilisers as hoof and horn, urea-formaldehyde, etc., can result in a temporary increase in the pH of the compost to the neutral point and above. Significant quantities of free ammonia can then be present.

Other forms of nitrogen can also be toxic to plants. Usually ammonium is converted to nitrites, which are then rapidly converted to nitrate nitrogen, and only very small or trace amounts of nitrite nitrogen will be found in normal soils. The *nitrobacter*, which converts NO_2 to NO_3, is killed by steam sterilisation and is also inhibited by the presence of high levels of ammonia. Consequently nitrites may accumulate to toxic levels for a short period in steamed soils (Fig. 5.5 p. 92). Court, Stephen and Waid (1962) show that, in addition to the risk of free ammonia toxicity occurring when urea was used as a nitrogen source, nitrites could also cause toxicity. Using a bioassay technique, they found that, when the nitrite nitrogen in the soil reached 110 ppm, the growth of young maize seedlings was reduced to 30% of that of the control plants.

The use of steam-air mixtures to heat soil to temperatures below 100°C has been found to produce less nitrite nitrogen than sterilisation with steam. White (1971) found that 42 days after soils had been heated to 100°, 71° or 60°C the NO_2–N values were 22, 14 and 13 ppm respectively.

Whilst ammonium, free ammonia and nitrite are the forms of nitrogen which are generally believed to be of most consequence following heat sterilisation, several workers have cited toxicities arising from other nitrogenous compounds.

SOLUBLE ORGANIC COMPOUNDS

Some soils release toxic amounts of soluble organic matter when they are heat treated. Walker and Thompson (1949) found that after steam sterilisation a peat soil gave an increase of 650% in the amount of soluble organic matter present, whereas another soil with a low organic matter content (loss on ignition 11·4%) gave only a 25% increase in the soluble organic matter following steaming. Schreiner and Shorey (1909) isolated dihydroxystearic acid from organic soils and found it injurious at all concentrations and lethal at 100 ppm.

MANGANESE

In addition to the increase in the amount of soluble organic matter and the various forms of nitrogen following the heat treatment of soils, the solubility of several minerals including Ca, Zn, K, Cu, Al and Mn is also increased; of these, manganese is usually the most important from the phytotoxicity point of view. Early workers noted the large increase in the soluble and exchangeable forms of manganese following steam sterilisation and this problem has received particular attention from workers at the Glasshouse Crops Research Institute. Here, steam sterilisation of the brickearth soil of the Hamble series, upon which the Institute is sited, usually results in manganese toxicity if no remedial action is taken.

Davis (1957) found that steam sterilisation of this brickearth soil (pH 4·26) resulted in an immediate increase in the exchangeable form of manganese, and 11 days after steaming the manganese level had risen from 106 to 1 444 ppm. The effect of soil pH on manganese release was studied by applying calcium carbonate at varying intervals before sterilisation. It was found that the longer the interval between liming and sterilising, the smaller was the increase in the amounts of water-soluble and exchangeable manganese produced. Increasing the soil pH from 5·1 to 7·0 reduced the amount of manganese in the tissue of young tomatoes from 4 900 ppm to 1 300 ppm, whilst the addition of superphosphate to the soil reduced the manganese level still further to 800 ppm.

Messing (1965) found that the sterilisation of soil at pH 5·9 produced a much smaller release of manganese than when the soil was at pH 5·3. Furthermore, below pH 5·3, the amount of extractable manganese, i.e. the water soluble plus the exchangeable manganese, increased during the period of the experiment, whereas above pH 5·9 the manganese levels decreased during the experiment. The effect on the uptake of manganese by plants of adding superphosphate to the steamed soil, was found to be dependent upon the pH of the soil. If the pH was high, then adding superphosphate invariably decreased the pH of the soil and increased the amount of manganese absorbed, but when the pH was low, applying superphosphate reduced the amount of manganese taken up by the plants. Where manganese toxicity is expected to occur following steam sterilisation, applying a solution of 0·6 g/l (0·1 oz./gall.) of monoammonium phosphate ($=150$ ppm P) will often give beneficial results.

For advisory purposes, Harrod (1971) has used the amount and forms of manganese present in the soil, either before or after sterilisation, to estimate the risk of manganese toxicity to plants (Table 12.3). It is assumed that some of the easily reducible manganese will be converted into the exchangeable form as a result of steam sterilisation. Most workers are agreed that there is little risk of manganese toxicity resulting from steam sterilisation providing that the soil pH is 6·0–6·5 before steaming. Page (1962) suggested that the observed relationship between soil pH and the forms of manganese results

Table 12.3. *Manganese toxicity risks.*

Risk	Manganese levels (ppm)	
	Before steaming. Exchangeable plus easily reducible forms	After steaming. Exchangeable form
Safe	100	75
Doubtful	100–250	75–100
Dangerous	250	100

from the formation of manganese complexes with the organic matter in the soil.

Several workers have found that, by comparison with steam which gives a soil temperature of 100°C, steam–air mixtures having lower temperatures give much lower rates of manganese release. Values reported by Dawson et al. (1965) for a range of steam–air mixtures are given in Table 12.4. At

Table 12.4. *Effect of the sterilising temperature on the release of manganese.*

Soil temperature		Water soluble plus extractable manganese (ppm)
Control	(unheated)	8
60°C	(140°F)	11
71°C	(160°F)	15
82°C	(180°F)	27
88°C	(190°F)	38
93°C	(200°F)	58
100°C	(212°F)	83

temperatures below 88°C, the manganese levels also showed a rapid decline; 10 days after the soil had been heated to 82°C, there were only 5 ppm of manganese present, whereas in soil heated to 100°C the corresponding value was 50 ppm.

ALUMINIUM

During the course of his investigations into the manganese toxicity of lettuce grown in steam sterilised soils, Messing found that aluminium could also be a related factor. Increases in the amount of extractable aluminium, after soil sterilisation, ranged from 15 to 100% of that found in the unsteamed soil. Those measures already recommended to control manganese toxicity, i.e. liming to pH 6–6·5, and applying superphosphate to the soil, will also control aluminium toxicity.

12.9 RULES FOR HEAT STERILISATION

Observance of the following rules when sterilising soils will ensure the maximum benefits with the minimum risks:

(1) Before a new and unknown soil is steam sterilised and used on a large scale, first determine its response to heat sterilisation and the risk of nitrogen and manganese toxicity by growing a trial amount of seedlings such as antirrhinum.

(2) If steam is used for sterilisation, ensure that both the soil and the steam are 'dry'. Wet soil and wet steam cause inefficient heating and leave the soil in a poor physical condition. Heat the soil rapidly (within 10 minutes), and, after allowing a further 10 minutes with the steam turned off, remove the soil and cool it quickly.
(3) Ensure that the soil pH is between 6·0 and 6·5, avoid soils with a high organic matter content and do not add large amounts of organic nitrogen in the base fertiliser. Early commencement of liquid feeding is better and safer than trying to give too much nitrogen in the base fertiliser.
(4) If the sterilised soil or the made up compost has to be stored before use, *it must be kept at a low temperature* to minimise biological changes which occur in the forms of nitrogen. *Do not store for more than three weeks.*
(5) Low-temperature steam–air mixtures give less risk of toxicity than high temperature steam, but the length of the heating period must be adjusted to the temperature of the steam-air mixture: 160°F for 30 minutes is recommended.

Heat sterilisation offers the grower the means of ensuring a pest and disease free compost with which to commence the propagation of seedlings and the growing of young pot plants. Providing that he has observed the above rules, sterilisation checks to plant growth will be minimal and the benefits will far exceed any disadvantages. The increasing popularity of composts made from materials which do not require sterilisation probably means that the peak of interest in soil sterilisation for pot plants has passed. Specific pest and disease problems will nevertheless continue to occur and it may well be necessary to resort to the heat sterilisation of these materials at times.

Chapter 13

Chemical sterilisation

By comparison with mineral soils, many of the materials used to make loamless composts can be considered already sterile as far as plant pathogens are concerned. Some materials, such as foam plastics, perlite and vermiculite have been subjected to very high temperatures during their manufacture and have virtually been sterilised already. Whilst many peats are free of pathogens, some samples may contain *Rhizoctonia*, and some bark samples may be infected with *Phytophthora cinnamomi*. The incidence of fungi in composts made from peat is, however, normally low, and it is common practice in Britain not to treat peat with heat or chemicals except when growing certain crops which are particularly susceptible to soil-borne diseases. For example, peat used for cucumber propagation is often heat sterilised or treated chemically against *Pythium* attack; antirrhinum is another plant which is susceptible to *Pythium*. Tomato, chrysanthemum and French marigolds are examples of plants which are not particularly susceptible to soil-borne fungal attacks.

Whenever there is the risk of pest or disease infestation in the bulky materials used to make composts and means of heat sterilisation are not available, chemical treatment can be considered as an alternative. Chemicals have the advantage of a low capital cost by comparison with steam and other forms of heat sterilisation. Their main disadvantage is that they often leave toxic residues, and treated soil must be left for a period before it can safely be used.

13.1 SOIL FUMIGANTS

The traditional materials for chemical sterilisation, such as carbon disulphide, cresylic acid and formaldehyde, are now seldom used. They have been replaced by new materials which are more effective, have a much wider spectrum of activity and can give results that are often comparable to heat sterilisation. Chemicals in this group include volatile liquids, powders and fumigants.

METHYL BROMIDE

This material is effective against most fungi, insects, nematodes and weed seeds. Experience in the USA suggests, however, that it does not give adequate control of *Verticillium*. It is an almost odourless and colourless gas at normal temperature and is highly toxic to humans; usually 2% of chloropicrin (tear gas) is added as a warning agent. In Britain, the material can be applied only by approved contractors and it is essential that the necessary safety precautions are fully observed and implemented.

The soil to be treated is enclosed in a polythene sheet and the gas, liquified under pressure, applied at the rate of 4 lb./100 cu.ft of soil (0·7 kg/cu.m) through a vapouriser. The soil should be moist and the temperature must not be less than 50°F. After treatment, the soil must remain sealed in the sheet for 4–5 days and then be freely ventilated for 4–10 days before it is used. The fumes are slightly toxic to some plants but the greatest risk is caused by the formation of inorganic bromides in the soil. Plants such as antirrhinum, carnation and salvia have been found to be particularly susceptible to bromide toxicity. The toxic residues formed in the soil can be removed by leaching but very heavy rates of water application are required. In the USA a mixture of methyl bromide with 33% chloropicrin, known as Dowfume MC-33, is available. This is more effective against *Verticillium*.

The dangers in using methyl bromide, requiring the employment of specialist operators, the problem of toxic residues and the difficulty in their removal from organic materials, all combine to discourage its use on materials to be used for making potting composts. Methyl bromide is, however, being increasingly used to treat glasshouse soils for tomato growing.

CHLOROPICRIN

Chloropicrin or tear gas is effective against most fungi and insects, but is not as effective as methyl bromide against nematodes. It is a heavy, almost colourless liquid which volatilises when injected into the soil. As the vapour is very pungent and can quickly cause nausea and tears, a respirator must be worn. It is injected into the soil at the rate of 3 ml/cu.ft and the soil must then remain covered with a gasproof cover for 24 hours. The soil must be in good physical condition to allow uniform penetration and must have a temperature of not less than 60°F. Chloropicrin fumes are toxic to living plants and all traces of the gas must be removed from the soil before it is used for making composts. The area chosen for treating the soil should also be well away from growing plants. After treatment, an interval of 2–4 weeks is required before the soil can be safely used. The higher the soil temperature and the smaller the heap, the shorter is the period required to remove the fumes.

In the USA, chloropicrin is available as 'Larvicide' and 'Picfume'.

METHYL ISOTHIOCYANATE (METHAM SODIUM)

This material has a general fungicidal, insecticidal and nematicidal action. Small quantities of potting soil can be treated by making up a solution of 2 pints of 33% metham sodium in 25 galls and applying this to the moist soil in 6 in. layers at the rate of 5 galls/cu.yd (1% solution at 30 l/cu.m). The metham sodium breaks down in the soil to form methyl isothiocyanate. In addition to the liquid formulation, which is available as 'Sistan' or 'Vapam', the material is also available in a prilled form under the name of 'Dazomet' (98% active ingredient). The prilled material is mixed uniformly with the moist soil at the rate of 6 oz./cu.yd (220 g/cu.m). After treatment, the soil is covered for three weeks and then turned three times at 14 day intervals. The vapour is very toxic to plants and it is essential that all traces of the methyl isothiocyanate have disappeared from the soil before it is used. This is readily tested by comparing the germination of cress seed in the treated and untreated soil.

In addition to the above-mentioned general purpose fumigants, two other materials are effective nematicides, i.e. dichloropropene-dichloropropane (DD) and ethylene dibromide (EDB).

All of the above chemicals kill the nitrifying bacteria in the soil, producing an accumulation of ammonium nitrogen. Whilst their effect on the nitrogen cycle is not as great as that produced by sterilisation with heat, the same precautions regarding toxicities should be followed. Gasser and Peachy (1964) have reported that Dazomet had a greater effect on retarding nitrification than metham sodium. Methyl bromide was also found to increase the rate of mineralisation of soil organic nitrogen more than the other sterilants.

13.2 FUNGICIDES

The low risk of fungal and insect attack in composts made from peat means that the peat is not usually treated with either heat or chemicals before it is used in compost making. There is, however, some risk of the peat containing *Rhizoctonia, Phytophthora, Pythium* or *Thielaviopsis,* and some bedding plants are especially liable to attack by these fungi. A number of fungicides can be used for prophylactic or preventative disease control purposes; they are either added to the compost during mixing or applied as a drench immediately after the seedlings have been pricked out.

QUINTOZENE

The active ingredient of this material is Pentachloronitrobenzene (PCNB) and it is known under the trade names 'Botrilex' in Britain and 'Brassicol' and 'Terraclor' in the USA. It is effective against *Rhizoctonia* and is mixed into the compost at the rate of 1·5 lb./cu.yd (890 g/cu.m) two or three days before planting

250   *Chemical Sterilisation*

DRAZOXOLON

This is available under the trade name 'MilCol' in Britain. It is effective against *Pythium* and is widely used by bedding plant growers, especially on antirrhinum seedlings. It is used at the rate of 5–10 fl.oz./100 galls of water (0·8 to 1·6 ml/l). Seed boxes are thoroughly soaked by applying 15 fl.oz. (¾ pint or 425 ml) per box at seed sowing and again at pricking out.

Whilst this material is safe to handle when diluted, the concentrate is very toxic and must be handled with care. A chemical analogue of this material with the common name of metazoxolon has a much lower mammalian toxicity and is likely to replace drazoxolon for this application.

ETRIDIAZOLE

This is the proposed common name for a fungicide based on 5-Ethoxy-3-trichloromethyl-1,2,4-thiodiazole. Trade names for fungicides containing this material are 'Truban', 'Terrazole' and 'Aaterra'. It is used to control *Pythium* and *Phytophthora* in pot and bedding plants, and is either applied as a drench, at 3–10 oz./100 galls (19–62 g/100 l), or mixed into the compost at 1½–3 oz./cu.yd (55–110 g/cu.m). Repeat applications of the drench may be given at 4–12 week intervals as required.

CHINOSOL

This is a fungicide and bactericide based on potassium hydroxyquinoline sulphate and is sold in Britain under the trade name 'Cryptonol'. This material contains 14% of active ingredient and is used against a wide range of fungi, including *Botrytis, Phoma, Pythium, Phytophthora*, and *Thielaviopsis*. It is applied as a 0·1% solution at seven-day intervals.

CHLOROTHALONIL

This is known as 'Daconil 2787' in Britain, and 'Bravo' in the USA. It is used to control *Rhizoctonia*.

FENAMINOSULF

Known as 'Dexon' in the USA, it is used at 8 oz. of wettable powder per 100 galls of water to control *Pythium*. It is unstable in light and must be used immediately after diluting.

CHESHUNT COMPOUND

Whilst this is one of the early fungicides, it still provides a cheap and effective means of *Pythium* control. It can be purchased ready mixed, or prepared by mixing 2 oz. of finely ground copper sulphate with 11 oz. of ammonium carbonate and then stored for at least 24 hours in a tightly stoppered glass jar. It is used at the rate of 1 oz. dissolved in a little hot water and made up to 2 galls.

In addition to the above-mentioned fungicides, several mixtures have been made, based on the active ingredients of two different materials. This usually results in fungicides, with a wider spectrum of activity, which are also more effective against a specific pathogen. Examples of such materials are:

TERRADACTYL
This is based on etridiazole and chlorothalonil. It is effective against *Pythium, Phytophthora, Rhizoctonia* and *Botrytis,* when used at 4 oz./cu.yd (150 g/cu.m).

BANROT
This material contains 25% of etridiazole, which is used to control *Pythium* and *Phytophthora,* and 15% of thiophanate-methyl, which is active against *Rhizoctonia, Fusarium, Verticillium* and *Thielaviopsis.* Thiophanate-methyl is the active ingredient of Zyban (USA) and Mildothane (Britain). Banrot has been used on a wide range of pot-grown ornamentals at 8 oz./100 galls (50 g/100 l). There is, at present, less information on its use on bedding plants; some plants such as Impatiens and petunia are damaged at the above rate.

Caution. The formulations of the various fungicides mentioned in this section and their percentage of active ingredients varies with the manufacturer. Because of this, precise directions on the strength of application cannot be given. *It is essential that the manufacturer's instructions are followed.* It is also advisable to test the materials first on a small scale. Where there is an acute disease problem and the reaction of the various plant species to the fungicides is unknown, it is preferable to use heat to sterilise the materials before making up the compost.

13.3 INSECTICIDES

Normally, the materials used for making composts are not likely to contain insect pests, and no treatment is required. If peat-based composts have been stored for a period before they are used, there is the possibility of an infection of sciarid fly (fungus gnats) occurring. These insects are attracted by decomposing organic matter and are particularly troublesome in composts containing organic sources of nitrogen, e.g. hoof and horn. As a preventative measure, $4\frac{1}{2}$ oz. of a wettable powder formulation of Diazinon can be mixed into each cubic yard of compost (170 g/cu.m). This is effective in controlling the fly for 4–6 weeks. Subsequent attacks can be controlled by applying either a Diazinon or a Parathion drench, providing that the crops concerned are not susceptible to these chemicals. Aphelandra and maidenhair fern are liable to be damaged by Diazinon and cyclamen is damaged by Parathion.

The larvae of the vine weevil, which sometimes attack cyclamen, can be controlled by mixing Aldrin dust into the compost. The dust (1·25% a.i.) is normally added at the rate of $2\frac{1}{2}$ lb./cu.yd (1·5 kg/cu.m).

Chapter 14

Plant containers

By far the greater proportion of the total amount of compost made will be used to grow plants in containers of one sort or another. The usual range of containers includes seed boxes made of wood or plastic and individual containers made of clay, plastic, cardboard, paper or peat. Only a relatively small amount of compost will be used to grow plants in borders or be compressed into blocks which have sufficient tensile strength not to require a retaining wall. The formulation and preparation of composts has so far been discussed without considering the effects which the type of container can have upon both the compost and the plant.

A review of the literature relating to plant containers has already been made (Bunt, 1960) and it is not now intended to attempt an overall evaluation of the types of container most suited for growing a particular crop. Rather, it is proposed to examine the interactions that can occur between the type of container, the compost and the plant.

The container traditionally used for the propagation of vegetable crops such as tomatoes and cucumbers, which are ultimately to be planted into their permanent cropping positions in the glasshouse border, has been the clay pot. This type of container has also been in general use for raising flower crops such as cineraria and gloxinia which remain permanently in the container. In Britain, the recent expansion of the plastics industry has resulted in the virtual replacement of the clay pot by pots made from various plastic materials, which are both cheaper and lighter to handle than clay pots. In some European countries and also in the USA the plastic pot has not yet achieved such a universal acceptance.

14.1 CLAY V. PLASTIC POTS

The essential difference between these two types of container is one of porosity. Plastic pots are non-porous, whilst clay pots can have varying degrees of porosity, depending upon the manner in which they are made. Porous clay pots differ from plastic pots in three principal ways:

(1) Water is lost by evaporation from the clay wall, hence the compost can be drier.
(2) The latent heat of evaporation from the clay wall results in a reduction in the temperature of the compost.
(3) The flow of water from the compost into the clay wall results in the loss of some nutrients to the plants.

WATER LOSS

The comparative water loss from clay and plastic pots can best be determined from fallow pots. This overcomes any effect which the pot may have upon the amount of growth made by the plant, which in turn affects the rate of water loss. Clay pots lose water by evaporation from the compost surface and also from the clay wall, whereas plastic pots lose water only by evaporation from the compost surface. The relative importance of water lost from these two sources was determined by Bunt and Kulwiec (1971) and was found to vary with the season (Table 14.1). In winter, the mean daily evaporative

Table 14.1. *Mean evaporative water loss from fallow plastic and clay pots (grams per day).*

Container type	Water loss
Winter	
Plastic	8·1
Clay	14·9
Clay wall only	6·3
Summer	
Plastic	22·8
Clay	34·4
Clay wall only	8·7

loss from 10·6 cm (4¼ in.) diameter plastic pots was approximately 8·1 g, and from comparable clay pots it was 14·9 g, a difference of 6·8 g or an increase of 85% by the clay pots. In summer, the mean daily losses were 22·8 g for plastic pots and 34·4 g for clay pots, a difference of 11·6 g or an increase of 50% by the clay pots. It was found that, in winter, approximately 42% of the total amount of water lost by evaporation from the clay pots occurred through the pot wall whilst in summer the loss through the clay wall was reduced to 25% of the total loss.

In practical terms this means that, whilst the *actual rate* of drying out of clay pots will always be greater than that of plastic pots, *the relative rate at which clay pots dry out will be greater in winter than it is in the summer*. Plants grown in plastic pots in winter will, therefore, be much more susceptible to over-watering and to waterlogged conditions than will those grown in clay pots. For this reason, it is of more importance to use a compost which drains freely and has a high air capacity when growing in plastic pots. The beneficial effect of using porous containers in winter, when growing under excessively wet conditions caused either by the compost having a poor physical structure with low aeration or by watering the pots too frequently is shown in Figure 14.1. Whilst it is impossible to separate entirely other

Fig. 14.1 Reduced growth in plastic pots resulting from over-watering. When the frequency of watering was reduced and the physical structure of the compost improved, plants grown in plastic pots were not inferior to those grown in clay pots.

factors such as compost temperature and salinity, it can be concluded from other treatments in the experiment that the effect shown in Figure 14.1 is predominantly due to excess water and lack of aeration. Primula and cyclamen are two examples of plants which are sensitive to over-watering and wet conditions in winter. It should be noted that, whilst the porous clay pot will allow the passage of water through the pot wall, it is not permeable to air. The improvement in compost aeration when using clay pots is solely due to the loss of water by evaporation and its replacement with air drawn down from the compost surface.

In winter, the water loss through the clay wall has been shown to be almost equal to that lost from the compost surface. This means that plants in clay pots will require more frequent watering than those in plastic pots. In the case of young seedlings which, because of their small leaf area, have a very low transpirational loss, plants in clay pots will require watering about twice as frequently as those in plastic pots. As the plants grow larger and the water loss by transpiration becomes greater than the loss by evaporation, the difference in the frequency of watering required by the two types of pot is reduced. In the case of mature pot plants with large leaf areas, there will be very little difference in the frequency with which the different containers require watering. The slightly higher soil moisture tensions that may develop in clay pots, because of their greater rate of evaporation, will have no significant effect on plant growth in winter when the rate of water demand by the plant is low. In summer, however, when there is a greater rate of water loss by transpiration, any increase in the soil moisture stress in the clay pot is more likely to check plant growth.

Richards (1974) has examined the effect of the number and size of the drainage holes on the amount of water retained in composts in plastic containers. He calculated that with only a minimal water pressure of 1 mm, an unobstructed drainage hole of 6·35 mm diameter would pass 1 732 ml of water per hour (i.e. a $\frac{1}{4}$ in. hole will pass three pints of water per hour). In practice, the drainage holes are not completely unobstructed and the rate at which water will pass through the hole will be somewhat less than this. The most effective position of the drainage hole was found to be in the base of the pot. When the same number of drainage holes were positioned around the side of the pot so that the lower edge of each hole was directly at the base, drainage was less effective. Pots which had 'crocks' placed in the base, i.e. pieces of plastic to serve as coarse drainage material (which was formerly the practice amongst private gardeners) actually retained more water than those pots without crocks. The crocks broke the continuity of the water film between the compost particles and the bench surface and so impeded the flow of water from the pot.

It was concluded that the number and size of holes in plastic pots are not normally factors which limit the drainage. The structure of the compost and the type of surface upon which the pots are stood, i.e. an open type bench or a sand or gravel surface, are the most important factors in allowing the water to drain from the pots (see also pp. 56 and 212).

TEMPERATURE

When water evaporates, heat is lost, the latent heat of water evaporation being 585 calories per gram at 20°C. The increased rate of evaporation from clay pots can therefore be expected to result in the temperature of the compost being lower than that of plastic pots. Compost temperatures in clay and plastic pots were measured under a wide range of environmental conditions by Bunt and Kulwiec (1970). The difference in temperature between the two types of container varied with the rate of water evaporation and ranged from about 1°C (1·8°F) by night in winter to 4°C (7·2°F) by day in summer.

Although the temperature difference of 1°C in winter is small and might therefore be expected to have little effect on plant growth, experiments have shown that, in winter, plants in plastic pots made about 15% more growth than those in clay pots. This was because the temperature of the compost was often below the optimal level for plant growth and it then became a limiting factor. It is not generally realised that under radiation conditions on a cold, clear winter's night, the temperature of the pots falls steadily during the night, and by dawn the compost temperature can be as much as 5·5°C (10°F) below that of the air temperature of the glasshouse; i.e. at a controlled air temperature of 60°F, the compost temperature can be only 50°F. A typical example of the type of container affecting plant growth by way of compost temperature is shown in Figure 14.2. In this experiment, both plants had been watered by

Fig. 14.2   In winter, plants in plastic pots make more growth than those in clay pots because of the higher soil temperature *providing that* there are no adverse effects of salinity and over-watering. In summer, the temperature in plastic pots may be too high and so restrict plant growth.

tensiometer readings and salinity levels were low; the difference in growth was due to the effect of the container on the compost temperature.

In summer, however, the temperature of the compost can be appreciably above the optimal value; temperatures of up to 42°C (108°F) have been recorded in experiments at the Glasshouse Crops Research Institute in midsummer. At such high temperatures, growth can be adversely affected and evaporative cooling becomes beneficial; plants in clay pots then make more growth than those in plastic pots. The temperature of the compost is also influenced by the colour of the pot; clay pots are terra-cotta in colour whereas plastic pots are available in a range of colours. Measurements made with thermocouples have shown that black and terra-cotta plastic pots have virtually the same compost temperature, whilst white plastic pots can be as much as 4°C (7·2°F) lower than the other colours. Often the daytime temperature of the compost in white plastic pots will be approximately the same as that in clay pots. This is because the lower rate of solar heat absorption by the white plastic pot is approximately equal to the evaporative cooling effect of the clay pot. By night, however, the colour of the pot has no effect on its temperature and the clay pot is then cooler than any of the plastic pots. Claims that pots made of expanded polystyrene have a higher temperature by night than pots made of high density polystyrene have not been confirmed by experimental measurements.

Another factor which influences the temperature of the compost is the type of bench on which the pots are stood. Bunt and Kulwiec (1970) also included in their studies on porous and non-porous pots the effect which solid and open surface benches had on the temperature of the pots. It was found that, by night, pots stood on benches with solid surfaces were cooler than pots stood on benches having open (weldmesh) surfaces. This advantage of the open type bench surface in maintaining higher pot temperatures varied with the climate outside the glasshouse. On mild nights, the difference in temperature between pots stood on the two bench surfaces was small, whilst on very cold nights, pots stood on open-surface benches were approximately 1°C (1·8°F) warmer than comparable pots stood on benches with solid surfaces Lack of air drainage from the bench causes compost temperatures to be lower on benches with a solid surface. On a cold night with a clear sky, the large loss of heat by radiation results in an inverted pool of cold air which forms over the surface of the bench, thereby lowering the temperature of the pots. On a bench with an open surface, this cold air is able to drain down through the bench on to the floor of the glasshouse where it is reheated and mixed into the convection currents rising from the heating pipes. On a dull, cold day in winter, pots on a solid bench will also be cooler than those on an open bench for the reason already stated. On a bright summer's day, however, pots stood on a solid bench will have a higher temperature than those stood on an open bench. This is because of the higher surface temperature of the solid bench and also the reflected radiation.

NUTRIENT ABSORPTION

The third important effect of the container concerns plant nutrition. Water that moves from the compost into the pot walls and is then lost by evaporation contains plant nutrients; these nutrients are deposited in the clay wall and are effectively lost to the plant. Often the build-up of nutrients in the pot wall is seen as a powdery white deposit on the outer surface of the wall. Measurement of the salinity of the compost in clay pots has shown there is a high nutrient concentration in the centre of the pot and a relatively low nutrient concentration in the compost adjacent to the inside wall. This loss of nutrients

258 Plant Containers

into the pot wall can be beneficial in the case of young seedlings grown in composts having high salinity values. Under these conditions, the clay pot has an advantage over the plastic pot. For example, if tomato seedlings have been pricked out in winter into a John Innes Potting Compost No. 3 instead of into a JIP-1 compost, then the seedlings grown in clay pots will become established and grow more quickly than those in plastic pots. Conversely, when the nutrient levels in the compost are not too high and the growth rate is rapid, any loss of nutrients into the pot wall is detrimental and plants grown in clay pots will show starvation symptoms and a reduced growth rate before comparable plants grown in plastic pots.

An example of the retarding effect which a high salinity has on the growth of plants in plastic pots is shown in Figure 14.3.

Fig. 14.3 Using a compost with a high salinity can cause reduced growth in plastic pots. Clay pots are able to absorb some of the salts and so give better results.

POT CLEANING

Both clay and plastic pots should be thoroughly cleaned before being re-used, and plastic pots, because of their smooth, non-absorbent surface, are more easily washed than clay pots. In addition to the removal of compost particles, all traces of fungi and bacteria which might infect subsequent crops should also be removed. Clay pots can be steam sterilised but most types of plastic pot will melt if heated to this temperature; polypropylene is one type of plastic that can be steamed.

Of the chemicals that have been tried as disinfectants, a formaldehyde soak for 10 min. at a strength of 1 pint in six galls (20 ml/l) of water has given the best results (Nichols and Jordon, 1972). Because of the irritating vapour, treatment should be given in the open, and rubber gloves worn. After treatment, the pots should be stacked and covered with a plastic sheet for 24 hrs, then uncovered and hosed down with clean water at intervals for a few days until the smell of formaldehyde has disappeared. If the pots are allowed to dry out too quickly, paraformaldehyde is formed and this can be phytotoxic.

14.2 PAPER AND PEAT POTS

Pots made of paper and compressed peat have the advantage that the plant need not be removed from the pot before planting and no labour is required in gathering up the empty pots and cleaning them before re-use.

Paper pots can be subdivided into two groups, those which have been treated to make them waterproof and those which have not. Waterproofing prevents the paper from absorbing water, thereby allowing it to retain its strength and to resist bacterial decomposition. Having a dry wall with no evaporative cooling effect, the compost temperature in this type of pot is similar to that in plastic pots. Untreated paper pots absorb water and compost temperature in them responds in a similar way to that in clay pots. Pots made of compressed peat, which are usually made with a binding agent to increase their tensile strength, absorb water and behave in a similar way to clay pots with respect to temperature and nutrient absorption. Plant roots are, however, able to penetrate the compressed peat wall.

One important factor regarding the use of pots made of paper, peat and similar materials, is their susceptibility to bacterial decomposition and the subsequent reduction in the amount of mineral nitrogen available to the plant. Some materials decompose much more readily than others and the temporary lock-up of mineral nitrogen can result in a reduced growth rate by comparison with plants grown in plastic or clay pots. Provided that this situation can be recognised at an early stage, it can be corrected with a nitrogen feed. Most pots made of compressed peat now have some mineral nitrogen added during their manufacture to allow for this phenomenon.

When plants in paper and peat pots are planted into the field, it is important to see that the surrounding soil is kept moist. If the paper is allowed to dry out, the roots may have some difficulty in penetrating the pot wall and growing into the soil; this can result in a severe check to plant growth.

Providing that the characteristics of the different containers are kept in mind and the management adjusted accordingly, they can all be used successfully for the propagation of tomatoes, cucumbers, etc. Where ornamental pot plants have to be transported before marketing and then kept in a dwelling house, obviously the container must have good tensile strength and adequate aesthetic appeal. For this reason plastic and clay pots are preferred for these crops.

Appendices

1 METRIC CONVERSIONS

	To convert	To	Multiply by
Mass	pounds	kilograms	0·4536
	ounces	grams	28·3495
	kilograms	pounds	2·2046
	grams	ounces	0·0353
Length	yards	metres	0·9144
	inches	centimetres	2·54
	metres	yards	1·0936
	centimetres	inches	0·3937
Area	square yards	square metres	0·8361
	square inches	square centimetres	6·4516
	square metres	square yards	1·1959
	square centimetres	square inches	0·1550
Pressure	pounds/in^2	atmospheres	0·0680
	pounds/in^2	kg/m^2	703·0696
	atmospheres	pounds/in^2	14·6959
	kg/m^2	pounds/in^2	0·0014
Volume and Capacity	cubic yards	cubic metres	0·7645
	cubic feet	litres	28·3168
	bushels	cubic metres	0·0364
	gallons (Imperial)	litres	4·5461
	gallons (US)	litres	3·7854
	fluid ounces (Imperial)	mililitres	28·4122
	fluid ounces (US)	mililitres	29·5727
	cubic metres	cubic yards	1·3080
	litres	cubic feet	0·0353
	cubic metres	bushels	27·4967
	litres	gallons (Imperial)	0·2199
	litres	gallons (US)	0·2642
	mililitres	fluid ounces (Imperial)	0·0352
	mililitres	fluid ounces (US)	0·0338

Appendices 261

Density and concentration	lb./yd^3	kg/metre^3	0·5932
	oz./yd^3	g/metre^3	37·0797
	oz./bushel	g/litre	0·7795
	oz./gall. (Imperial)	g/litre	6·2361
	oz./gall. (US)	g/litre	7·4891
	kg/metre^3	lb./yd^3	1·6855
	g/metre^3	oz./yd^3	0·0269
	g/litre	oz./bushel	1·2829
	g/litre	oz./gall. (Imperial)	0·1603
	g/litre	oz./gall. (US)	0·1335
Energy	British thermal unit	kilojoules	1·0551
	BTU/lb.	joules/kg	2 326
	BTU/ft^3	kilojoules/m^3	37·2589
	BTU/ft^3 °F	kilojoules/m^3 °C	67·0661
	kilowatt hours	megajoules	3·6

2 IMPERIAL AND US CAPACITY MEASURES

1 Imperial gall.	=	1·2009 US galls
1 Imperial gall.	=	4·5459 litre
1 Imperial gall.	= 160	fl.oz.
1 Imperial fl.oz.	=	0·9607 US fl. oz.
1 US gall.	=	0·8326 Imperial galls
1 US gall.	=	3·7853 litre
1 US gall.	= 128	fl. oz.
1 US fl. oz.	=	1·0408 Imperial fl. oz.
1 Imperial bushel	=	1·0321 US bushels
1 US bushel	=	0·9689 Imperial bushels

3 ILLUMINATION AND SOLAR RADIATION UNITS

1 lumen ft^{-2}	= 1 foot candle
1 lumen m^{-2}	= 1 lux
1 foot candle	= 10·76 lux
1 calorie cm^{-2} min^{-1}	= 697·6 joules m^{-2} sec^{-1}
	= 697·6 watts m^{-2}
*7 500 lumens ft^{-2}	= approx. 1 calorie cm^{-2} min^{-1}

* The relationship of lumens to calories changes slightly with the time of the year and the amount of cloud. The figure of 7 500 lumens per calorie is an average value.

4 ATOMIC WEIGHTS

Hydrogen	H	1·00	Sulphur	S	32·066
Boron	B	10·82	Chlorine	Cl	35·457
Carbon	C	12·01	Potassium	K	39·096
Nitrogen	N	14·008	Calcium	Ca	40·08
Oxygen	O	16·00	Manganese	Mn	54·93
Sodium	Na	22·997	Iron	Fe	55·85
Magnesium	Mg	24·32	Copper	Cu	63·54
Aluminium	Al	26·97	Zinc	Zn	65·38
Phosphorus	P	30·98	Molybdenum	Mo	95·95

5 FORMULAE AND MOLECULAR WEIGHTS OF SOME COMMONLY USED CHEMICALS

Aluminium sulphate	$Al_2(SO_4)_3 \cdot 18H_2O$	666
Ammonium nitrate	NH_4NO_3	80
Diammonium phosphate	$(NH_4)_2HPO_4$	132
Monoammonium phosphate	$NH_4H_2PO_4$	115
Ammonium sulphate	$(NH_4)_2SO_4$	132
Borax	$Na_2B_4O_7 \cdot 10H_2O$	382
Solubor	$Na_2B_8O_{13} \cdot 4H_2O$	412
Boric acid	H_3BO_3	62
Calcium carbonate	$CaCO_3$	100
Calcium hydroxide	$Ca(OH)_2$	74
Calcium oxide	CaO	56
Calcium nitrate	$Ca(NO_3)_2$	164
Copper sulphate	$CuSO_4 \cdot 5H_2O$	249
Ferrous sulphate	$FeSO_4 \cdot 7H_2O$	278
Magnesium sulphate	$MgSO_4 \cdot 7H_2O$	246
Manganese sulphate	$MnSO_4 \cdot 7H_2O$	277
Potassium chloride	KCl	74
Potassium nitrate	KNO_3	101
Potassium sulphate	K_2SO_4	174
Urea	$CO(NH_2)_2$	60

6 CHEMICAL GRAVIMETRIC CONVERSIONS

To convert	To	Multiply by
NH_4	N	0·821
NO_3	N	0·226
N	NH_4	1·285
N	NO_3	4·427
P_2O_5	P	0·436
PO_4	P	0·326
P	P_2O_5	2·291
P	PO_4	3·066
K_2O	K	0·830
K	K_2O	1·205
$CaCO_3$	Ca	0·400
CaO	Ca	0·714
Ca	$CaCO_3$	2·497
Ca	CaO	1·399
$MgCO_3$	Mg	0·288
MgO	Mg	0·603
Mg	$MgCO_3$	3·467
Mg	MgO	1·657

7 TEMPERATURE CONVERSIONS

To convert °F to °C : $(°F - 32) \times \frac{5}{9}$
To convert °C to °F : $(°C \times \frac{9}{5}) + 32$

°F	°C	°F	°C
32	0	75	23·9
40	4·4	80	26·7
45	7·2	100	37·8
50	10·0	120	48·9
55	12·8	140	60·0
60	15·6	160	71·1
65	18·3	180	82·2
70	21·2	200	93·3

An increase of 1°F = 0·555°C
An increase of 1°C = 1·8°F

Bibliography

Aaron, J. R. 1974. Personal communication.
Adams, R. P. 1973. Personal communication.
Aldrich, R. A., J. W. White and P. E. Nelson 1972. Aerated steam. II: Engineering requirements, design and operation of systems for aerated steam treatment of soil and soil mixtures. *Pennsylvania Flo. Gro. Bull.* **253**, 3–7.
Allen, R. C. 1943. Influence of aluminium on the flower colour of hydrangea macrophylla D.C. *Contributions from Boyce Thompson Institute* **13**, 221–42.
Allison, F. E. 1965. *Decomposition of wood and bark sawdusts in soil, nitrogen requirements and effects on plants.* US Dept. Agric. Tech. Bull. No. 1332.
Allison, F. E., R. M. Murphy and C. J. Klein 1963. Nitrogen requirements for the decomposition of various kinds of finely ground woods in soils. *Soil Sci.* **96**, 187–91.
Alvey, N. G. 1955. Adapting John Innes composts to grow *Ericas. J. Roy. Hort. Soc.* **80**, 376–81.
  1961. Soil for John Innes composts. *J. Hort. Sci.* **36**, 228–40.
Anon, 1954. *Capillary watering of plants in containers.* Min. Agric. Fish and Food, Hortic. Machinery Leaflet No. 10, London: HMSO.
Arnold Bik, R. 1970. Nitrogen, salinity, substrates and growth of Gloxinia and Chrysanthemum. *Mededeling No. 3, Overdruk Van: Verslagen landbouwkundige Onderzoekingen* 739.
  1972. Influence of nitrogen, phosphorus and potassium rates on the mineral composition of the leaves of the Azalea variety Ambrosius. *Colloquium Proceedings no.* **2**, 99–102. *Potassium Institute.*
  1973. Personal communication.
Arnon, D. I. and C. M. Johnson 1942. Influence of hydrogen ion concentration on the growth of higher plants under controlled conditions. *Plant Physiol.* **17**, 525–39.
Asen, S. and C. E. Wildon 1953. Nutritional requirements of greenhouse chrysanthemums growing in peat and sand. *Quart. Bull. Mich. Agric. Exp. Sta.* **36**, 24–9.
Baker, K. F., ed. 1957. *The UC system for producing healthy container-grown plants.* Calif. Agric. Exp. Sta. Manual **23**.
  1967. Some microbiological effects of soil treatment with steam and chemicals. *Proc. Washington State University's Greenhouse Growers Institute* (June), 20–2.
Barry, T. A. 1969. Origins and distribution of peat types in the bogs of Ireland. *Irish Forestry* **26** (2), 1–14.
Batson, F. 1972. Azalea. In the *Ball Red Book,* 12th ed., 195–214. Chicago: G. T. Ball Inc.
Bennett, A. C. and F. Adams 1970. Concentration of NH_3 (aq.) required for incipient NH_3 toxicity to seedlings. *Soil Sci. Soc. Amer. Proc.* **34**, 259–63.

Bernstein, L. 1963. Osmotic adjustment of plants to saline media. II: Dynamic phase. *Amer. J. of Bot.* **50**, 360–70.

Bingham, F. T. 1959. *Micro-nutrient content of phosphorus fertilisers. Soil Sci.* **88**, 7–10.

Bingham, F. T. and M. J. Garber 1960. Solubility and availability of micro-nutrients in relation to phosphorus fertilisation. *Soil Sci. Soc. Amer. Proc.* **24**, 209–13.

Blomme, R. and T. G. Piens 1969. Kunstmatige bodems en bemesting van Azalea [Artificial soils and manuring of Azaleas]. *BVO mededelingen Bedrijfsvoorlichtinchtingsdienst voor de Tuinbouw in de Provincie Oost.* Vlaanderen VZWO No. 51.

Bollen, W. B. 1953. Mulches and soil conditions, carbon and nitrogen in farm and forest products. *J. Agric. Food Chem.* **1**, 379–81.

1969. Properties of tree barks in relation to their agricultural utilisation. Pacific N.W. Forest and Range Exp. Sta. Portland Oregon, USDA. *Forest Service Research Paper* PNW-77.

Boodley, J. W. and R. Sheldrake Jr 1972. *Cornell Peat-Lite mixes for commercial plant growing.* N.Y. College of Agric., Cornell Univ., New York. Information Bulletin no. 43.

Branson, R. L., R. H. Sciaroni and J. M. Rible 1968. Magnesium deficiency in cut-flower chrysanthemums. *Calif. Agric.* **22**, No. 8, 13–14.

Bunt, A. C. 1954*a*. Steam pressure in soil sterilisation. I: In bins. *J. Hort. Sci.* **29**, 89–97.

1954*b*. Steam sterilisation. Steam–air mixture. *Ann. Rep. John Innes Horticultural Institute,* p. 28.

1956. An examination of the factors contributing to the pH of the John Innes Composts. *J. Hort. Sci.* **31**, 258–71,

1960. A review of the literature on plant containers and moulded blocks with special reference to the porosity of pots. *Rep. Glasshouse Crops Res. Inst.* 1959, 116–25.

1961. Some physical properties of pot plant composts and their effect on plant growth. II: Air capacity of substrates. *Plant and Soil* **15**, 13–24.

1963. The John Innes Composts: Some effects of increasing the base fertiliser concentration on the growth and composition of the tomato. *Plant and Soil* **19**, 153–65.

1971. The use of peat-sand substrates for pot chrysanthemum culture. *Acta Horticulturae* **18**, 66–74.

1972. The use of fritted trace elements in peat-sand substrates. *Acta Horticulturae* **26**, 129–40.

1973*a*. Losmless substrates for pot plants. Micro-element problems. *Rep. Glasshouse Crops Res. Inst.* 1972, 66–7.

1973*b*. Factors contributing to the delay in the flowering of pot chrysanthemums grown in peat-sand substrates. *Acta Horticulturae* **31**, 163–72.

1974*a*. Physical and chemical characteristics of loamless pot-plant substrates and their relation to plant growth. *Proc. Symposium Artificial Media in Horticulture,* 1973, 1954–65. Ghent: Int. Soc. Hort. Sci.

1974*b*. Loamless substrates for pot plants. Micro-element supply. *Rep. Glasshouse Crops Res. Inst.* 1973, p. 78.

Bunt, A. C. and P. Adams 1966*a*. Some critical comparison of peat–sand and loam-based composts with special reference to the interpretation of physical and chemical analysis. *Plant and Soil* **24**, 213–21.

1966*b*. Loamless composts. *Rep. Glasshouse Crops Res. Inst.* 1965, 119–20.

Bunt, A. C. and Z. J. Kulwiec 1970. The effect of container porosity on root environment and plant growth. I: Temperature. *Plant and Soil* **32**, 65–80.

1971. The effect of container porosity on root environment and plant growth. II: Water relations. *Plant and Soil*, **35**, 1–16.
Bylov, V. N., N. V. Vasilyevskaya and L. P. Vavilova 1971. Physiochemical properties of vermiculite and its use in floriculture. *Byull. Gl. Bot. Sada.* **80**, 59–63.
Cameron Brown, C. A. and P. Wakeford 1947. *Electrical soil sterilisation by immersion heaters.* British Electrical and Allied Industries Research Association Technical Report W/T14.
Cappaert, I., O. Verdonck and M. De Boodt 1974. Barkwaste as a growing medium for plants. *Proc. Symposium Artificial Media in Horticulture* 1973, 2013–22. Ghent: Int. Soc. Hort. Sci.
Colgrave, M. S. and A. N. Roberts 1956. Growth of the azalea as influenced by ammonium and nitrate nitrogen. *Proc. Amer. Soc. Hort. Sci.* **68**, 522–36.
Court, M. N., R. C. Stephen and J. S. Waid 1962. Nitrite toxicity arising from the use of urea as a fertiliser. *Nature* **194**, 1263–5.
Criley, R. A. and W. H. Carlson 1970. Tissue analysis standards for various floricultural crops. *Florists' Review* **146**, no. 3771, 19–20, 70–3.
Dänhardt, W. and G. Kühle 1959. [Experiments on the most favourable peat-clay ratio in standard peat soils to be used for pot plants.] (In German.) *Arch. Gartenb.* **7**, 157–74.
Davies, J. N. 1957. Steam sterilisation studies. *Rep. Glasshouse Crops Res. Inst.* 1954/55, 70–9.
Dawson, J. R., R. A. H. Johnson, P. Adams and F. T. Last 1965. Influence of steam-air mixtures, when used for heating soil, on biological and chemical properties that affect seedling growth. *Ann. Appl. Biol.* **56**, 243–51.
De Boodt, M. and O. Verdonck 1971. Physical properties of peat and peat-moulds improved by perlite and foam plastics in relation to ornamental plant-growth. *Acta Horticulturae* **18**, 9–27.
1972. The physical properties of the substrates in horticulture. *Acta Horticulturae* **26**, 37–44.
Dempster, C. D. 1958. Clay dust compost solves the loam problem. *Commercial Grower* **3245**, 569–71.
Eaton, F. M. 1941. Water uptake and root growth as influenced by inequalities in the concentration of the substrate. *Plant Physiol.* **61**, 545–64.
1942. Toxicity and accumulation of chloride and sulfate salts in plants. *J. Agr. Res.* **64**, 357–99.
Farnham, R. S. 1969. Classification system for commercial peat. *Proc. 3rd Int. Peat Congress, Quebec,* 85–90. Ottawa: Dept. Energy, Mines and Resources.
Fisher, R. A. 1926. The arrangement of field experiments. *J. Min. Agric.* (London) **33**, 503–13.
Fruhstorfer, A. 1952. *Soil mixture for horticulture.* Complete specification, Pat. Spec. 670, 907. London: Brit. Patent Office.
Furuta, T., R. H. Sciaroni and J. R. Breece 1967. Sulphur-coated urea fertiliser for controlled release on container-grown ornamentals. *Calif. Agric.* **21**, (9) 4–5.
Gabriels, R., H. Engles and J. G. Van Onsem 1972. Nutritional requirements of young azaleas grown in peat and coniferous litter. *Symposium International Peat Society* (Helsinki).
Gammon, N. 1957. Root growth responses to soil pH adjustments made with carbonates of calcium, sodium or potassium. *Proc. Soil Crop Sci. Soc. Florida* **17**, 249–54.
Gartner, J. B., S. M. Still and J. E. Klett 1973. The use of hardwood bark as a growth medium. *The Int. Plant Prop. Society, Combined Proc.* **23**, 222–31.

Gasser, J. K. R. 1964. Effects of solutions of urea and ammonium and potassium salts on the germination of kale, barley and wheat. *Chemical Industry*, **40**, 1687–8.

Gasser, J. K. R. and J. E. Peachey 1964. A note on the effects of some soil sterilants on the mineralisation and nitrification of soil nitrogen. *J. Sci. Fd. Agric.* **15**, 142–6.

Gauch, H. G. and C. H. Wadleigh 1944. Effects of high salt concentrations on growth of bean plants. *Bot. Gaz.* **105**, 379–87.

Green, J. L. 1968. Perlite – advantages and limitations as a growth medium. *Colo. Flo. Gro. Ass. Bull.* **214**, 4–8.

Guttormsen, G. 1969. Accumulation of salts in the sub-irrigation of pot plants. *Plant and Soil* **31**, 425–38.

Harrod, M. F. 1971. Metal toxicities in glasshouse crops. A discussion of problems encountered in advisory work on soils of pH 6·0 and above. In *Trace Elements in Soils and Crops*, Tech. Bull. **21**, 176–92. London: HMSO.

Hatfield, J. D., A. V. Slack, G. L. Crow and H. B. Shaffer 1958. Corrosion of metals by liquid-mixed fertilisers. *J. Agric. Food Chem.* **6**, 524–31.

Hillman, W. A. and H. B. Posner 1971. Ammonium ion and the flowering of *Lemna perpusilla*. *Plant Physiol.* **47**, 586–7.

Holden, E. R., N. R. Page and J. I. Wear 1962. Properties and use of micro-nutrient glasses in crop production. *J. Agric. Food Chem.* **10**, 188–92.

Homes, M. V. 1963. The method of systematic variations. *Soil Sci.* **96**, 380–6.

Hooper, L. J. 1973. Personal communication.

Johnson, J. 1919. The influence of heated soils on seed germination and plant growth. *Soil Sci.* **7**, 1–104.

Johnson, P. 1968. *Horticultural and agricultural uses of sawdust and soil amendments*. Paul Johnson, 1904 Cleveland Ave., National City Calif. 29 050, USA.

Joiner, J. N. and C. A. Conover 1965. Characteristics affecting desirability of various media components for production of container-grown plants. *Proc. Soil and Crop Sci. Soc. Florida* **25**, 320–8.

Joiner, J. N. and W. E. Waters 1973. The influence of cultural conditions on the chemical composition of six tropical foliage plants. *Florida Foliage Grower* **10**, No. 8, 1–2.

Jungk, A. 1968. Influence of nitrogen and potassium concentration on nutrient yield. *Proc. 6th Colloquium Int. Potash Inst.* 310–19.

Kersten, M. S. 1949. Thermal properties of soils. *Minn. Univ. Engr. Exp. Sta. Bull.* **28**, 1–228.

Kipliner, D. C. and H. K. Tayama 1970. Foliar analysis information for floral crops. *Ohio State Florists' Assn. Bull.* **493**, 2–9.

Kivinen, E. and V. Puustjärvi 1972. Classification of peat. *4th Int. Peat Congress* **5**, 97–103.

Kohl, G. C., A. M. Kofranek and O. R. Lunt 1955. Effect of various ions and total salt concentrations on Saintpaulia. *Proc. Amer. Soc. Hort. Sci.* **68**, 545–50.

Laurie, A. 1931. The use of washed sand as a substitute for soil in greenhouse culture. *Proc. Amer. Soc. Hort. Sci.* **28**, 427–31.

Lawrence, W. J. C. and J. Newell 1939. *Seed and Potting Composts*. London: George Allen & Unwin.

Long, M. I. E. and G. W. Winsor 1960. Isolation of some urea-formaldehyde compounds and their decomposition in soil. *J. Sci. Fd Agric.* **11**, 441–5.

Lucas, R. E. and J. K. Davis 1961. Relationships between pH values of organic soils and availabilities of 12 plant nutrients. *Soil Sci.* **92**, 177–82.

Maynard, D. N. and A. V. Barker 1969. Studies on the tolerance of plants to ammonium nutrition. *J. Amer. Soc. Hort. Sci.* **94**, 235–9.

Messing, J. H. L. 1965. The effects of lime and superphosphate on manganese toxicity in steam-sterilised soil. *Plant and Soil* **23**, 1–16.
Mitchell, R. L. 1954. Trace elements in Scottish peats. *Int. Peat Symposium, Section B3*. Dublin.
  1971. Trace elements in soils. In *Trace elements in Soils and Crops*, Tech. Bull. **21**, 8–20. London: HMSO.
Morris, L. G. 1954. *The steam sterilising of soils*. National Inst. Agric. Engr. Report No. 24.
Morriston, T. M., D. C. McDonald and J. A. Sutton 1960. Plant growth in expanded perlite. *N.Z. J. Agric. Res.* **3**, 592–7.
Needham, P. 1973. Private communication.
Nelson, L. B. 1965. Advances in Fertilisers, In *Advances in Agronomy* **17**, 1–84. New York: Academic Press.
Nelson, P. V. 1972. *Greenhouse Media. The use of Cofuna, Floramull, Pinebark and Styromull*. North Carolina Agric. Exp. Sta. Tech. Bull. No. 206.
Nelson, P. V. and J. W. Boodley 1966. Classification of carnation cultivars according to foliar nutrient content. *Proc. Amer. Soc. Hort. Sci.* **89**, 620–5.
Nelson, P. V. and K. H. Hsiek 1971. Ammonium toxicity in chrysanthemum: critical level and symptoms. *Comm. in Soil Science and Plant Analysis* **2**, 439–48.
Nichols, L. P. and M. H. Jodon 1972. Chemical soaks for prevention of growth of pathogenic organisms on clay and plastic pots. *Penn. State Flo. Gro. Bull.* **250**, 1, 6–7.
North, C. P. and A. Wallace 1959. Nitrogen effects on chlorosis in macadamia. *Calif. Macadamia Soc. Yearbook* **5**, 54–67.
Oertli, J. J. and O. R. Lunt 1962. Controlled release of fertiliser materials by incapsulating membranes. I: Factors influencing the rate of release. *Soil Sci. Soc. Amer. Proc.* **26**, 579–87.
Ogg, W. G. 1939. Peat. *Chemistry and Industry* **58**, 375–9.
Olsen, S. R. 1953. Inorganic phosphorus in alkaline and calcareous soils. *Agronomy* **4**, 89–122.
Owen, O. 1948. The occurrence and correction of magnesium deficiency in *Solanum capsicastrum* under commercial conditions. *Rep. Exp. Res. Sta. Cheshunt*, 80–1.
Page, E. R. 1962. Studies in soil and plant manganese. II: the relationship of soil pH to manganese availability. *Plant and Soil* **16**, 247–57.
Patterson, J. B. E. 1971. Metal toxicities arising from industry. In *Trace elements in soils and crops*. Tech. Bull. **21**,193–207. London: HMSO.
Pearson, H. E. 1949. Effect of waters of different quality on some ornamental plants. *Proc. Amer. Soc. Hort. Sci.* **53**, 532–42.
Penningsfeld, F. 1962. *Die Ernährung Im Blumen- und Zierplanzenbau*. Berlin/Hamburg: Verlag Paul Parey.
Pickering, S. U. 1908. Action of heat and antiseptics on soils. *J. Agric. Sci.* **3**, 32–54.
Post, K. 1946. Automatic watering. *New York State Flo. Gro. Bull.* **7**, 3–8.
Prasad, M. and P. A. Gallagher 1972. Sulphur coated urea, casein and other slow-release nitrogen fertilisers for tomato production. *Acta Horticulturae* **26**, 165–73.
Puustjärvi, V. 1968. Cation exchange capacity in sphagnum mosses and its effect on nutrient and water absorption. *Peat and Plant News* **4**, 54–8.
  1969. Basin-Peat culture. *Peat and Plant News* **2**, 20–4.
  1970a. Mobilisation of nitrogen in peat culture. *Peat and Plant News* **3**, 35–42.
  1970b. Degree of decomposition *Peat and Plant News* **4**, 48–52.
  1973. Personal communication.
Rader, L. F., L. M. White and C. W. Wittaker 1943. The salt index – a measure of the effect of fertilisers on the concentration of the soil solution. *Soil Sci.* **55**, 201–18.

Read, P. E. and R. Sheldrake Jr 1966. Correction of chlorosis in plants grown in Cornell Peat-Lite mixes. *Proc. Amer. Soc. Hort. Sci.* **88**, 576–81.

Reeker, R. 1960. The prevention of molybdenum deficiency in growth media from peat. *Soils and Fertilisers* **23**, abstract no. 2869.

Richards, L. A., ed., 1954. *Diagnosis and improvement of saline and alkaline soils.* United States Dept. of Agriculture, Agricultural Handbook No. 60.

Robertson, R. A. 1971. Nature and extent of Scottish peat resources. *Acta. Agralia Fennica* **123**, 233–41.

Rutland, R. B. 1972. Correlation of the electrical conductivities of the saturated paste extract (EC_e) and the 1:2 soil-to-water extract (EC_e) as a function of saturation percentage in greenhouse soil mixes. *Hort. Science* **7**, 190–2.

Schreiner, O. and E. C. Shorey. 1909. The isolation of harmful organic substances from soils. *USDA, Bureau of Soils, Bull.* **53**, 5–53.

Seeley, J. G. 1947. Automatic watering of potted plants. *New York State Flo. Gro. Bull.* **23**, 1–9.

Sheard, G. F. 1940. An investigation of the partial sterilisation of soil for horticultural purposes with special reference to the use of electrical sterilising equipment. MSc. Thesis, University of Leeds.

Sheldrake, R. Jr and O. A. Matkin 1971. Wetting agents for peat moss. *Acta Horticulturae* **18**, 37–42.

Smilde, K. W. 1971. Evaluation of fritted trace elements on peat substrates. Institute for Soil Fertility, Groningen, Rapport **1**.

Stoffert, G. 1973. The concept of work systems as an aid to rationalizing production of ornamental plants. *Acta Horticulturae* **31**, 99–106.

Swaine, D. J. 1962. *Trace element content of fertilisers.* Commonwealth Bureau of Soils Technical Communication No. 52.

Thompson, L. M. 1957. *Soils and soil fertility.* New York: McGraw-Hill.

Tod, H. 1956. High calcium or high pH? A study of the effect of soil alkalinity on the growth of Rhododendron. *J. Scottish Rock Garden Club* **V**, 1–8.

Wadleigh, C. H. and A. D. Ayers 1945. Growth and biochemical composition of bean plants as conditioned by soil moisture tension and salt concentration. *Plant Physiol.* **20**, 106–32.

Walker, T. W. and R. Thompson 1949. Some observations on the chemical changes effected by the steam sterilisation of glasshouse soils. *J. Hort. Sci.* **25**, 19–35.

Wall, R. F. and F. B. Cross 1943. *Greenhouse studies on the toxicities of Oklahoma salt contaminated waters.* Okla. Agric. Exp. Sta. Tech. Bull. **T-20**.

Wallace, A. 1962. *A decade of synthetic chelating agents in inorganic plant nutrition.* Arthur Wallace, 2278 Parnell Avenue, Los Angeles 64, California, USA.

Wallace, A. and O. R. Lunt 1960. Iron chlorosis in horticultural plants. A review. *Proc. Amer. Soc. Hort. Sci.* **75**, 819–41.

Walsh, T. and T. A. Barry 1958. The chemical composition of some Irish peats. *Proc. Royal Irish Academy,* **59**, section B, 305–28.

Warren Wilson, J. and J. Tunny 1965. Defects of perlite as a medium for plant growth. *Aust. J. Exp. Agric. and Animal Husbandry* **5**, 137–40.

Waters, W. E., W. Llewellyn and J. NeSmith 1970. The chemical, physical and salinity characteristics of twenty-seven soil media. *Proc. Flo. Sta. Hort. Soc.* **83**, 482–8.

Waters, W. E., J. NeSmith, C. M. Geraldson and S. S. Woltz 1972. The interpretation of soluble salt tests and soil analysis by different procedures. *Florida Foliage Grower* **9**, no. 6, 1–10.

Watson, W. 1913. Soils suitable for azalea culture. In *The Gardener's Assistant, Vol. 1,* 150–4. London: The Gresham Publishing Co.

Wells, D. A. and R. Soffe 1962. A bench method for the automatic watering by capillarity of plants grown in pots. *J. Agric. Engineering Res.* **7**, 42–6.

White, J. W. 1971. Interaction of nitrogenous fertilisers and steam on soil, chemicals and carnation growth. *J. Amer. Soc. Hort. Sci.* **96**, 134–7.

1974. Dillon Research Fund, Progress report on research at Penn. State. *Penn. Flo. Gro. Bull.* **273**, 3–4.

White, J. W. and J. W. Mastalerz 1966. Soil moisture as related to container capacity. *Proc. Amer. Soc. Hort. Sci.* **89**, 758–65.

Whitt, D. M. and L. D. Baver 1930. Particle size in relation to base exchange capacity and hydration properties of Putnam clay. *J. Amer. Soc. Agron.* **29**, 703–8.

Wind, G. P. 1968. *Root growth in acid soils*. The Netherlands Inst. for Land and Water Management Research, Technical Bulletin 55.

Winsor, G. W. and M. I. E. Long 1956. Mineralisation of the nitrogen in urea-formaldehyde compounds in relation to soil pH. *J. Sci. Fd. Agric.* **7**, 560–4.

Winsor, G. W., J. N. Davies and D. M. Massey 1963. Salinity studies. I: Effect of calcium sulphate on the correlation between plant growth and electrical conductivity of soil extracts. *J. Sci. Fd. Agric.* **14**, 42–8.

Index

Acidity (*see also* pH) 70–4; bark 33; effect of fertilisers 74, 168, 195; peat 73–4
Air capacity 43, 46, 52–8, 165; and irrigation systems 211–12; effect of container depth 52–3; requirement of plants 46, 58, 254
Aldrin 251
Algicides, use in irrigation systems 211
Aluminium 142–3; availability and pH 142–3; hydrangea flower colour 107, 143; phosphate 63, 109, 111, 143, 191, 207; sulphate, rate of use 143; toxicity 245
Amino acids 91, 116, 122
Ammonia, free 16, 87–8, 92–3, 101, 111, 142; toxicity 88, 90, 93, 101, 157
Ammonifying bacteria 16, 88, 91–2, 220, 242
Ammonium: fixation by vermiculite 38; pH effect 71, 86, 92–3, 99–100, 132, 225; toxicity 86–7, 99, 220, 227, 242
Ammonium polyphosphates 113, 180, 192
Analytical: solvents 35, 64, 131; weight and volume/basis comparison 45, 62, 65, 103
Anions 61, 63–4, 130; exchange capacity 38, 42, 63, 216
Antagonism, between elements 68, 86, 96, 108, 119, 122, 140–1
Available water 43, 53, 58, 165
Azalea composts 87, 166–9

Bacteria: ammonifying 88, 91–2, 133, 248; nitrifying 16, 38, 87–9, 92, 133, 220, 242
Bactericides, uses in irrigation systems 207–8
Baking 241
Banrot 251
Bark 30–4; acidity 33; bulk density 33, 42; cation exchange capacity 33; composts 59, 164–6; decomposition rate 30–1, 165; lime requirement 33; mineral composition 32; nitrogen requirement 30–1, 164–5; toxicities 30, 33–4
Basin culture, peat 150, 152
Bench type: effect on compost temperature 257; effect on container drainage 212, 255

Blockages, capillary irrigation 191–2, 207–8
Blossom-end rot, tomatoes 86
Boiler feed water, treatment 234–5
Boron 131–5, 148; availability and pH 36, 72, 144; content of fertilisers 134–5, 144, 147; deficiency and nitrogen source 101, 132–3; deficiency symptoms 37, 132; fertilisers 134–5, 144, 181; fritted form 134, 144–5; in liquid feeds 134–5; requirement of composts, 134; toxicity 107, 130, 134, 144, 149, 157
British Thermal Unit 231, 261
Bulk density 36, 43–5, 56, 65; bark 33, 42; peat 27, 42, 45, 123; peat-sand mixtures 45, 56; perlite 38, 42; plastics, 40–2; sand 37, 42; vermiculite 38, 42

Calcium 116–19; deficiency symptoms 116–17
Calcium fertilisers 117–19; calcic limestone 120; calcium carbonate 118, 121, 126, 127; calcium hydroxide 118, 126; calcium nitrate 89–90, 99, 121, 123, 126–7, 133, 181, 187–9; calcium oxide 118, 126; calcium sulphate 110, 119, 121, 126, 127; neutralising values 118; pH control 117
Calcium sulphate, and compost salinity 49, 80, 122, 127
Calculation of liquid feeds: Imperial measure 184–7; metric measure 184–7; US measure 184–9
Capillary irrigation systems 50, 208–13; salinity effect 213
Capillary water, in composts 43
Carbon: Nitrogen ratio 30–2, 101, 103–6; nitrogen requirements 30–2, 165; peats 103–6, 123; shredded bark 30, 32, 165; straw 30, 32; use of sugar solutions 101
Cation exchange capacity 19, 60–3, 65; bark 33; loam 62, 216; peat 19, 21, 27, 62; perlite 39; sawdust 33; vermiculite 38, 62
Cations 42, 161, 130
Chalk 118

271

Chelated micro elements 96, 136, 140–1, 145–6, 154–5; in liquid feeds 112–13, 140–1, 146, 162, 168, 180
Chemical analysis: solvents 35, 64, 131; weight v volume presentation 45, 62, 65, 103
Chemical sterilisers 247–51; chloropicrin 248; methyl bromide 248; methyl isothiocyanate (metham sodium) 249
Cheshunt compound 250
Chinosol 250
Chloride 141–2; in water supply 142, 204; toxicity 116, 142, 204
Chlorine free, 142
Chloropicrin 40, 248
Chlorothalonil 250
Chromium toxicity 35
Citric acid, use in irrigation systems 208
Clay 37; cation exchange capacity 37, 62; content of loam 217–18; powdered, in composts 37,150
Clay pots, characteristics 252–9
Coffee waste 29
Cofuna 34
Compaction 58, 234
Comparison of peat and mineral soil 23–5, 64
Compost formulation, loamless 150–69; Belgium 166–7; Denmark 151–2; Finland 152–3; Germany 153–4; Ireland 154–5; Netherlands 155, 167; United Kingdom 155–7; United States 158–66, 168–9
Compost preparation and mixing 51, 169–73, 221
Compost salinity: and irrigation systems 177–9; and water content 49–50; 177–9, 212
Compost storage 99–103, 156–9, 225–7, 251; ammonium toxicity 99–101, 157; boron deficiency 101–2; nitrogen transformations 99–101, 169; pH effect 99–100
Composts, loam (John Innes) 215–27; calcifuge plants 227–8
Composts, loamless: advantages and disadvantages 17–18; formulations 151–69; moisture content and mixing 51; storage 99–103, 156
Conductivity, interpretation of readings 79–81
Conductivity meter 168, 192–4
Container capacity, water content 43, 46, 52–3
Container cleaning 258–9
Container colour, and temperature 256–7
Container irrigation systems 205–14
Container porosity: nutrient absorption 257–8; temperature effect 255–7; water loss 46, 253–5
Container types: clay v plastic 211, 252–9; paper and peat 259
Copper 69, 135–6, 147–8; availability and pH 72, 136; availability in peats 136; deficiency symptoms 135; requirement of composts 136; toxicity 36, 136, 146, 149
Corrosion: by liquid fertilisers 194, 208; in steam mains 234–5
Cycocel 23

Desorption curve 43, 48, 212
Diazinon 251
Diluting equipment 197–203; conductivity meter 168, 192–4; displacement diluters 198–9, 208; micropumps 202; positive injection diluters 201–2, 208
Dolomite limestone 120
Dolomitic limestone 120
Drainage, from composts 46, 52
Drainage holes, in containers 255
Drazoxolon 250
Drip irrigation systems 205–8
Dyes, in liquid feeds 192–3

Easily available water 43, 56–8
EDTA 145–6; in liquid feeds 146, 192, 207
EDDHA 146
Electrical conductivity 78–81; and osmotic tension 79; control of liquid feed strength 182, 187, 193, 201; of various liquid feeds 182, 189–91
Electrical soil sterilisers: electrode type 240–1; immersion heater type 239–40
Enmag 96; rate of use 157
Essential elements 82, 107, 130
Ethylene in soils 45
Etridiazole 250
Evapotranspiration 45, 211; seasonal changes 179

Fertiliser: acidifying effect 86, 93, 168; analysis 126, 147; analysis, element—oxide conversions 263; availability index, urea formaldehyde 93; dissolution rate, coated urea 97; dyes in liquid feeds 192–3; micro element content 147; solubilities 181, 192; use in liquid feeds 180–197
Fertilisers: calcium, 117–19; calcium carbonate 118, 121, 126–7; calcium hydroxide 118, 126; calcium oxide 118, 126; calcium sulphate 110, 119, 121, 126–7
Fertilisers, magnesium 76, 120–2; Dolomite limestone 120–1, 127; Dolomitic limestone 120; Kieserite 120, 126–7; magnesium carbonate 120, 126; magnesium chloride 76, 119; magnesium limestone 120–1; magnesium nitrate 119; magnesium oxide 120, 127; magnesium sulphate 76, 119–20, 122, 126
Ferilisers, nitrogen 90–1; ammonium nitrate 90, 126–7, 181; ammonium sulphate 89–90,

Index

126–7, 181, 195–6; calcium-ammonium nitrate 90–1; calcium nitrate 89–90, 99, 121, 123, 126–7, 133, 181, 187–9; urea 89–90, 120, 126–7, 133, 181

Fertilisers, phosphorus 110–14, basic slag, 113, 126; calcium metaphosphate 114, 126; diammonium phosphate 112, 126–7, 181; magnesium ammonium phosphate 95–6, 98, 113–14; monoammonium phosphate 112, 126–7, 181, 191–2; phosphoric acid 112, 126, 194–5; polyphosphates 113, 126, 192; potassium metaphosphate 114, 126; superphosphate 110–12, 126–7; superphosphoric acid 112,126

Fertilisers, potassium 115–16; Kainit 116; potassium carbonate 116, 126, 181; potassium chloride 116, 126, 181; potassium frit 116; potassium metaphosphate 114, 126; potassium nitrate 91, 116, 126, 181, 192; potassium sulphate 115–16, 126–7, 181

Fertilisers, slow release 91–8, 113–14; and steam sterilisation 98; basic slag 113, 126; calcium metaphosphate 114, 126; coated fertilisers 96–8; crotonylidene-diurea 95, 98; dried blood 91–3; hoof and horn 89, 91–4, 98–9, 123–5, 133; isobutyridene-diurea 95, 98; magnesium ammonium phosphates 95–6, 98, 113–14; oxamide 96; potassium frit 116; potassium metaphosphate 114, 126; rock phosphate 113; sulphur-coated urea 97–8; urea-formaldehyde 93–5, 98, 155, 157, 220, 226, 242

Field capacity, water content 46, 52
Flame pasteuriser 238–9
Flow rate control 199–201
Foliar sprays, micro elements 135–6, 138, 141, 146
Formaldehyde, pot cleaning 258–9
Fritted trace elements 134, 138, 143–5, 156–7, 160–1; availability and pH 144
Fumigants, soil 247–9
Fungi, thermal deathpoints 229–30
Fungicides, soil 249–51
Fungus gnat, control 251

Galvanised metal, toxicity 36, 141
Gravel 36
Grit 15, 55–6, 220
Growth regulators, and peat type 22–5
Gypsum 119

Heat capacity of soils 232–4; effect of water content 233
Heat conduction in soil 234, 239
Heat requirement, sterilisation 232–4, 241
Hoof and horn: 91–3, 99; ammonium toxicity 220–1; effect on pH 92–3, 99, 132, 242;

mineralisation rate 93, 98–9, 225; stored composts 91–4, 98–9, 225–7, 251

Ion exchange resins 42, 65
Immobilisation of nitrogen: C:N ratios 101, 103–6; in peats 105–6
Iron 138–41; availability and pH 72, 145–6, 148; chelates 96, 145–6; deficiency symptoms 138–40; frits 144; in peat 140; interactions in plant nutrition 87, 139–41, 148; supplying fertilisers 140–1
Iron deficiency, causes 87, 96, 139–40
Iron phosphate 63, 109, 111, 140, 148, 191, 207
Irrigated sand benches 210–11
Irrigation and water quality 42, 122, 139, 141–2, 188–9, 191, 194–6, 203–4; pH effects 188, 194–6; phosphorus precipitation 112; salinity effects 189–90
Irrigation systems 205–14; blockage prevention 207–8; capillary 50, 208–13; drip 205–8; flooded benches 213–14; mats 211; salinity effect 177–9, 196, 206–7, 213; sand 208–11

John Innes base fertiliser 216
John Innes composts 15–16, 23–5, 215–28; characteristics and use 64, 221–7; formulation, calcifuge plants, 227–8; formulation, potting composts 216; formulation, seed composts 216; seasonal effects on growth 224–5; storage 225–7; loam selection and preparation 16, 216–21

Kieserite 120, 126

Leaching, loss of nutrients 64–7, 175, 179, 206, 212
Leafmould 15, 215
Lime induced chlorosis 139
Lime requirement 73, 117; of bark 33; of loam 216–17; of peat types 73–4, 153, 157
Liming materials 117–19; calcium carbonate (chalk), 118 121; calcium hydroxide 118; calcium oxide 118; calcium sulphate 110, 119, 121, 126–7; Dolomite limestone 120–1, 127; Dolomitic limestone 120; magnesium limestone 120
Liming materials, neutralising values 118
Liquid fertiliser equipment 197–203; conductivity meter 78–9, 189, 193, 201; corrosion prevention 194; displacement diluters 198–9, 208; filters 207; flow rate control 199–201; micro pumps 202; positive injection diluters 201–2, 208; quality control 182, 187, 193–4, 201
Liquid fertilisers: acidity and pH control 168, 187, 194–6; base v liquid fertilisers 175–7;

boron containing 134–5; dyes 192–3, 198; formulation and calculations 180–9; magnesium 192, 197; molybdenum 138, 197; phosphorus 191–2; season and strength 177–9, 196–7; solubility rates 181, 192; specific gravity 184, 199; use of chelates 146, 180, 192

Liquid fertilisers, formulae: Imperial measure 184–7; metric measure 184–7; United States measure 184–9

Liquid fertilisers, phosphorus 191–2; factors causing precipitation 191–2, 207–8; grades of fertiliser 112, 191–2

Liquid fertilisers, salinity, 127, 182, 189–90; effect on compost 177–9, 213; effect of water 189; of various fertilisers 178, 182, 190

Loam: composts 215–28; lime requirement 217; preparation for composts 219–21

Loam selection 16, 216–19; aggregate stability 218; clay content 217–18; organic matter content 218; pH 216–17

Loam sterilisation 16, 218, 220; manganese toxicity 16, 217, 243–5; methods 231–41, 247–51; nitrogen toxicity and pH 217; steam requirement 232–4

Loam v loamless composts 18, 83, 131; analytical comparisons 65; availability of nutrients 64; nutrient loss by leaching 65–7

Macro elements 107–22; availability in loam v loamless composts 63–7; availability in peat 72; content of peats 83; content of plants 60, 82, 108, 114, 119, 128–9; function in plants 82, 108, 115–16, 119; interactions 108, 119–20, 122; interaction with season 80, 87, 123–5, 151, 175–9, 224–5

Mag Amp, rate of use 157

Magnesium 119–22; deficiency symptoms 119; foliar sprays 119; in liquid feeds 192; potassium antagonism 68, 96, 108, 119, 122

Magnesium ammonium phosphates 95–6, 98, 113–14; iron induced deficiency 96; potassium induced deficiency 96; rate of use 157; slow release nitrogen source 95–6; 98

Magnesium fertilisers 76, 120; Dolomite limestone 120–1, 127; Dolomitic limestone 120; Kieserite 120, 126–7; magnesium carbonate 120, 126; magnesium chloride 76, 119; magnesium limestone 120–1; magnesium nitrate 119; magnesium oxide 120, 127; magnesium sulphate 76, 119–20, 122, 126; toxicity of 76

Manganese 136–7, 148–9; availability and pH 72, 136, 148; deficiency symptoms 136; fertilisers 136, 147, 180; forms in soils 46

Manganese toxicity 16, 45, 137, 146, 149, 180, 243–5; and monoammonium phosphate 244; heat sterilisation 16, 243–5; in peats 137; pH effect 16, 217, 244

Matric tension 44, 46, 48, 50, 76; effect on plants 45, 49, 50, 76–7, 206, 213; effect on osmotic tension 49–50; units of measurement 46, 48–9

Methyl bromide 40, 248

Methyl isothiocyanate (metham sodium) 249

Micro elements 130–49; availability and compost storage 101–2; availability and pH 71–3, 130–1, 148; content of peats 131; content of plants 149; foliar sprays 135, 136, 138, 141; function in plants 131–2, 135–7, 139, 142; in chelated form 140–1, 145–6, 180; in fertilisers 147, 154; in fritted form 143–5, 156–7, 160–1; interactions 136, 139–40, 145–6, 148; release from sewage waste 35–6; toxicities 35, 130, 134, 136–7, 141

Mineral soil, peat comparisons 45, 122–3

Mineralisation of organic nitrogen: and ammonium toxicity 16, 91–3, 99–100, 220–1, 242; bacteria 16, 88, 92, 220; effect on micro element availability 101–2; effect on pH 71, 92–3, 99–100, 132, 225; factors controlling rate 93, 124; stages in 91–3

Mix formulations: Azalea 166–9; Denmark 151–2; Finland 152–3; Germany 153–4; Ireland 154–5; Netherlands 155; United Kingdom 155–7; United States 158–66

Mixing composts 169–73, 221; equipment 170–3; moisture content of peat 51, 169

Moisture factor, in salinity determinations 79

Molybdenum: and nitrogen source 137–8; availability and pH 72, 131; content of plants 137; deficiency symptoms 137; fertilisers 137–8; in liquid feeds 138

Monoammonium phosphate 112, 191–2; and manganese toxicity 244; grade for liquid feeds 191–2

Mortar rubble 15, 215

Nickel toxicity 35

Nitrate nitrogen 86–90; and iron deficiency 87, 139–40; and storage of composts 99–101, 156; calcium nitrate in liquid feeds 187–8; fertilisers 90–1; losses by leaching 65–7

Nitric acid 194, 203, 208

Nitrifying bacteria 16, 87–8, 92, 105; and pH 92; and steam sterilisation 16, 88, 220, 242; inhibition by ammonium nitrogen 89, 243

Nitrite toxicity 87, 89, 243; and urea 89; reduced by steam air mixtures 243

Nitrogen 82–106; acidifying effect 86, 93; ammonium toxicity 86–7; ammonium v.

nitrate forms 86–90, 156–7; content of peats 83, 103; deficiency symptoms 82, immobilisation 30–1, 38, 101, 103, 105, 259; loss by leaching 65–7; nitrite toxicity 89, 90; requirement and C:N ratios 30–2, 103; requirement of pot plants 83–6, 175–7; transformations in soils 88, 91–3, 105
Nitrogen forms 86–9, 121; and boron deficiency 93, 101, 132–3; and compost storage 99–101, 156–7; and iron deficiency 87, 139–40; and toxicities 16, 86–9, 99–101, 123–4, 220–2, 242–3; light intensity and carbohydrates 86–7; molybdenum requirement 138; plant preference for 86–9, 121, 123–4, 166; solubility in peats 64–7
Nitrogen, organic 91–5; and heat sterilisation 220–1, 242–3; mineralisation 91–3, 124
Nitrogenous fertilisers, mineral 90–1; ammonium nitrate 90, 126–7, 181, 195–6; ammonium sulphate 90, 126–7; calcium ammonium nitrate 90–1; calcium nitrate 90, 99, 121, 123, 126–7, 181, 187, 189; urea 89–90, 121, 126–7, 133, 181
Nitrogenous fertilisers, slow release 91–8; and steam sterilisation 98; coated fertilisers, 96–8; crotonylidene diurea 95, 98; dried blood 91–3; hoof and horn 89, 91–4, 98–9, 123–5, 133; isobutyridene—diurea 95, 98; magnesium ammonium phosphates 95–6, 98, 113–14; oxamide 96; sulphur coated urea 97–8; urea-formaldehyde 93–5, 98, 155, 157, 220, 226, 242
Non-ionic wetting agents 51–2
Nutrient availability: loam v loamless composts 64, 83; loss by leaching 65–7, 84
Nutrient interactions 68, 86, 96, 108, 111, 119–21, 123–4, 133, 136, 139–40, 145–6, 148, 220–1; with season 80, 87, 124–5, 138, 151, 175–9, 224–5
Nutrient uptake by plants 67–9, 84–6, 175–7

Organic matter: C:N ratios 30–4, 103; nitrogen immobilisation 30–2, 101, 103; toxicities 30, 35
Organic matter, nitrogen release 91–3; by chemical sterilisation 249; by heat sterilisation 242–3; from peats 83
Organic matter, types used in composts: bark 29–34; cofuna 34–5; leafmould 15; peat 19–29; pine needles 29, 166; sawdust 29–34; sewage waste 35–6
Osmocote, rate of use 160
Osmotic tensions in compost 44, 79, 85, 177–9; calcium sulphate effect 49, 80, 127; effect of compost water content 49–50, 58, 177–9; effect on plants 49, 58, 76–7; effect on water availability 76–7; methods of determination 78–81; plant tolerance 76–7, 79–81, 142, 168
Overwatering risk 40, 46, 57, 101, 139, 207, 211–12, 253–4; and air capacity 57, 166
Oxygen 45–6, 68

Pathogens: chemical control 247–51; heat control 17, 229–31, 235–7
Peanut hulls 29
Peat 16–17, 19, 219; bulk density 27; carbon: nitrogen ratio 103; cation exchange capacity 19, 21, 27, 62; composition, mineral 32, 83, 103 109, 115, 123, 131, 136; composition organic 26, 123; decomposition, von Post scale 22, 26–7; formation 19–20; lime requirement 73–4; loose volume, bale 27; nitrogen content 27, 83, 103; particle size 27, 55, 219; pH 27, 73, 131; production methods 28–9; properties of sphagnum peat 27; sterilisation 17; water content, in bale 27; water content, for mixing 51, 169; wetting 51–2, 160
Peat bogs, types 20
Peat, classification systems: Great Britain 20–1; International 22; United States 21–22
Peat-sand ratios, effects on physical properties 54–8
Peat types, botanical classification 21–2; sedge 21, 23–6, 42, 73–4, 103–5; sphagnum 21, 23–6, 42, 73–4, 103–5, 131
Perlite 17, 19, 38–9, 229; aluminium toxicity 39; cation exchange capacity 39; in composts 39, 59, 160–1; water availability 39
Permanent wilting percentage 44, 47
pF curve 44, 48–9
pH (acidity) 70–4; control by liquid feeds 168, 187, 194–6; definition 70; effect of fertilisers 71, 74, 86, 92, 111, 118, 194–6; effect of heat sterilisation 225, 242; effect of water quality 157, 194–6; free ammonia 92–3; interaction with nitrogen sources 94, 96, 110, 120–1, 123–4; lime requirement 73–4; 217; methods of measurement 70–1; nutrient availability 63, 71–3, 109, 130
Phosfon 23–5
Phosphoric acid 38, 112–13, 194, 203; and blockage of irrigation lines 112, 208; thermal grade material 112, 191, 195; wet process material 112, 195
Phosphorus 108–14; availability in peats 64; beneficial effects in composts 110–11, 124, 128; content of plant tissue 60, 108; deficiency symptoms 108–9; fixation 63–7, 109; forms and compost pH 109–11; leaching from composts 65–7; minor element interactions 148
Phosphorus fertilisers, for base application:

basic slag 113; calcium metaphosphate 114, 126; magnesium ammonium phosphate 95–6, 98, 113–14; potassium metaphosphate 114, 126; rock phosphates 113; superphosphates 110–12, 126–7
Phosphorus fertilisers, for liquid feeds 191–2; diammonium phosphate, 112, 126–7, 181; monoammonium phosphate 112, 126–7, 181, 191–2; phosphoric acid 112, 126, 194–5; polyammonium phosphate 113, 126; superphosphoric acid 112, 126
Pine needles 29, 166
Plant tissue analysis 60, 84, 119, 128–9, 149, 167, 176–7, 224
Plastic pots, characteristics 252–9
Plastics, in composts 16, 40, 42, 166; polystyrene 40; polyurethane 41; urea-formaldehyde 41
Polyammonium phosphates 113, 180, 192
Pore space 43–4, 46, 54; air capacity 46, 52–8; size distribution 39, 46, 53; water retention 46, 56
Pot cleaning 258–9
Pot types: clay 55, 211, 252–9; paper and peat 259; plastic 55, 211, 252–9
Potassium 114–16; content of plants 60, 114; deficiency symptoms 115; in mineral soils 115; in peats 115; loss by leaching 65–7
Potassium fertilisers 115–16; potassium carbonate 116, 126, 181; potassium chloride 116, 126–7, 181; potassium frit 116; potassium metaphosphate 114, 126; potassium nitrate 91, 116, 126–7, 181, 192; potassium sulphate 115–16, 126–7, 181
Precipitation in liquid feeds 191–2, 207; control measures 112, 114, 180, 192, 207–8
Pumice 39, 42

Quintozene 249

Rice hulls 29, 42
Rockwool 39, 151

Salinity 74–81; and compost water content 49–50, 177–9, 212; and osmotic concentration 79, 177; and water availability 76; control of liquid fertilisers 182, 187, 189–91; effect of container type 257–8; effect on plants 58, 76–8, 168, 226–7, 257–8; methods of measurement 78–81; moisture factor 79; non-electrolytes 80–1, 85, 190–1; of fertilisers 85, 127, 182, 189–91; of water supply 42, 142, 203–4; plant tolerance to 76–7, 79–81, 142, 153–4, 168, 203, 226–7
Salt index, fertilisers 127–8
Sand, particle size 36, 55, 219; for capillary benches 210; for composts 36, 55, 219; quality 36–7, 220
Sawdust 30–4, 42, 103; compost formulae 163–4
Sciarid fly, control 251
Sewage waste 35–6
Shredded bark 30–4; in composts 164–6
Sodium bicarbonate 195, 203
Soil fumigants 247–9
Soil fungicides 249–51
Soil solution 67, 75, 88, 109, 126–7
Soil temperature: and toxicities 242, 245; baking 241; container colour 256–7; container porosity 255–7; electrical sterilisation 239–41; flame pasteurisation 238; steam sterilisation 231–4; steam-air sterilisation 236–7
Soil water: container capacity 43, 53; field capacity 52; tension curve 44, 47
Soil: water ratio effect 71, 80
Solubility of fertilisers 181
Soluble organic compounds and toxicities 243
Solute stress 44, 49, 58, 76
Specific gravity 44, 184, 199
Specific heat: of soils 232, 236; of water 232
Spent hops 15, 215
Steam: dryness of 232, 234; latent heat 232; mains insulation 232; pressure in soil 232–3; requirement, soil sterilisation 232–4; sensible heat 231–2; separator 170, 232, 234
Steam-air mixtures 170, 235–7; biological aspects 235; chemical aspects 237, 243; temperatures of 236
Steam pressure, and heat content 232; and volume 232
Steam sterilisation of loam: biological changes 16, 220, 225–7, 235, 237, 242–3; chemical changes 16, 88, 220–1, 225, 242–5
Steam sterilisation toxicity: aluminium 245; ammonium 220, 225; manganese 243–5; soil temperatures 242, 245; soluble organic matter 243
Sterilisation methods: baking 241; chemical 247–51; electrical 239–41; flame pasteuriser 238–9; steam 231–4; steam-air 235–7
Sterilisation, thermal deathpoints 229–31; temperature: time relationship 230–1, 246
Sterilisation, toxicities 220, 225, 242–5
Stone wool 39, 151
Sterilisation equipment 169–70, 232, 234–5, 237
Storage of composts 99–103, 156–7, 169, 225–7, 251; nitrogen forms and transformations 99–101, 225–7; toxicities 101, 157
Sugar solution, and nitrogen toxicity 101
Sulphates and salinity measurements 80
Sulphur 31, 122; and pH reduction 227–8;

content of peats 122–3
Sulphur coated urea 97–8; dissolution rate 97
Sulphuric acid 194, 203
Superphosphate 63; beneficial effects of 110–11, 124; effect on compost pH 74, 109, 111; interaction with nitrogen forms 110; interaction with season 110, 124; micro element content 147; solubility and leaching 64–7
Surfactants 51–2, 160

Temperature, container colour effect 256–7; container porosity effect 252–8; steam sterilisation 231–2; steam–air mixtures 236–7
Tensiometer 47, 50
Terradactyl 251
Thermal deathpoint 229–31; temperature × time relation 230–1
Thermal efficiency, soil sterilisation 232–4
Tissue analysis 60, 84, 119, 128–9, 149, 167, 176–7, 224
Total porosity 55, 58
Total soil moisture stress 44, 46, 58, 76; matric and osmotic tension interaction 49–50, 76–7
Toxicity aluminium 245; ammonia 88, 99, 101, 242; ammonium 86–7, 99, 220–1, 242; bark and sawdust 30, 33; boron 107, 130, 134, 144, 149, 157; chromium 35; chlorides 122, 142, 204; copper 36, 136, 146, 149; free formaldehyde 41; magnesium salts 76; manganese 137, 146, 149, 180; molybdenum 149; nickel 35; nitrites 89, 243; sewage waste 35–6; soluble organic matter 33, 243; sulphates 122; surfactants 52; zinc 35, 141, 146, 149
Trickle irrigation systems 205–8

Ultra violet, water sterilisation 204
United States, liquid fertilisers 187–9; liquid fertiliser calculations 184–6; measures 187, 261

Urea, ammonia toxicity 88–9, 101, 110–11, 121; nitrite toxicity 89; non-electrolyte 81, 85; pH effect 89, 99, 110, 121, 133
Urea-crotonaldehyde 95, 98
Urea-formaldehyde, fertiliser 32, 93, 220, 226, 242; effect of pH 94; mineralisation rate 94, 98, 220; rate of use in composts 155, 157
Urea–formaldehyde, foam 41

Vapour pressure deficit 76–7
Vermiculite 16, 17, 19, 37–8, 59, 107, 115, 141, 151, 159–61, 229; ammonium fixation by 38; cation exchange capacity 38, 62; density 38; nutrient loss by leaching 66–7; pH 38; phosphorus fixation by 67
Vine weevil 251
Volume weight 27, 44
von Post scale 22

Water: absorption by composts 37, 40–1, 50; absorption by plants 50, 76–7; availability and salts 49, 75, 77; capacity of composts 44; flow rate control equipment 199–201; purification 204; quality for irrigation 42, 122, 139, 141, 188–91, 194–6, 203–4; regulations for diluting equipment 202; requirement, seasonal changes 45, 179
Water in composts 46–53; and air capacity 53; and container depth 52–3; comparison of irrigation systems 211–12; container capacity 43, 46, 79; effect on solute stress 49–50; 177–9; energy concept 46; ethylene production 45; field capacity 46; tension curve 47–8; tension equivalents 48–9
Waterlogged soils 46; iron deficiency 139; manganese toxicity 45
Wetting agents 51–2, 160
Wood ashes 15

Zinc 141, 147–9; deficiency symptoms 141; fertilisers 141; toxities 35, 141